New Studies in Biology

Subtidal Ecology

Elizabeth M. Wood
BSc, PhD
Marine Biologist, Basingstoke, Hampshire

CAMBRIDGE
UNIVERSITY PRESS

CAMBRIDGE UNIVERSITY PRESS
Cambridge, New York, Melbourne, Madrid, Cape Town, Singapore,
São Paulo, Delhi, Dubai, Tokyo

Cambridge University Press
The Edinburgh Building, Cambridge CB2 8RU, UK

Published in the United States of America by Cambridge University Press, New York

www.cambridge.org
Information on this title: www.cambridge.org/9780521427906

First published by Edward Arnold (Publishers) Ltd 1987
Re-issued in this digitally printed version by Cambridge University Press 2010

A catalogue record for this publication is available from the British Library

ISBN 978-0-521-42790-6 Paperback

General Preface to the Series

Recent advances in biology have made it increasingly difficult for both students and teachers to keep abreast of all the new developments in so wide-ranging a subject. The New Studies in Biology, originating from an initiative of the Institute of Biology, are published to facilitate resolution of this problem. Each text provides a synthesis of a field and gives the reader an authoritative overview of the subject without unnecessary detail.

The Studies series originated 20 years ago but its vigour has been maintained by the regular production of new editions and the introduction of additional titles as new themes become clearly identified. It is appropriate for the New Studies in their refined format to appear at a time when the public at large has become conscious of the beneficial applications of knowledge from the whole spectrum from molecular to environmental biology. The new series is set to provide as great a boon to the new generation of students as the original series did to their fathers.

1986

Institute of Biology
20 Queensberry Place
London SW7 2DZ

Preface

The ecology of the seashore around the British Isles, that unique zone of transition between land and sea, has been studied in great detail and described in many publications. The sublittoral shelf beyond it, although considerably more extensive, has received less attention, partly because of the practical difficulties of working in submerged habitats. However, it is an area rich in natural resources of significant economic and biological value, and is also of considerable educational and recreational interest to a growing number of people. I hope that this book will be a stimulus not only to all those who venture down and into the sea, but also those who prefer to view it from dry land.

1987   E.M.W.

Contents

1

Introduction

By the nineteenth century, interest in the natural history of the seashore around the British Isles had been awakened, and explorations were also being made into coastal and offshore waters. Nets had long been used in open water for the capture of fish, but it was not until 1828, when a fine-mesh tow net was first used by amateur naturalist J. Vaughan Thompson, that the study of planktonic organisms really began.

At around this time, the piece of equipment most readily available for studying the benthos was a simple dredge consisting of a bag of coarse netting held open on a rectangular iron frame. Edward Forbes began dredging in British waters in 1840, and his pioneering work was taken up with great enthusiasm by other marine naturalists and scientists. By the end of the 19th century several marine laboratories had been established, and as time progressed, sampling techniques became more refined, with emphasis being laid on collecting quantitative data. This was possible with the study of planktonic organisms and the biota of soft substrata, but was less easy to achieve in rocky areas.

The first detailed *in situ* observations in the shallow sublittoral zone of the British Isles were made using standard (helmet) diving apparatus. The best known study is that by Kitching and his associates of a subtidal gully at Wembury in Devon (Kitching *et al.*, 1934). The cumbersome gear used at that time was far from ideal, and it was not until the invention of the aqualung that shallow coastal waters became more easily and directly accessible to man. Even then, in the late 1950s, diving was carried out more for the sense of adventure than for any other reason. However, it was not long before the potential of the equipment was realized and serious biological work began. Now, *in situ* studies using the aqualung are recognized as an invaluable and essential part of many marine biological investigations, especially in rocky areas where work using more conventional techniques is not easy. Even so, the use of diving as a method of study has obvious limitations, not least of which are the need to avoid long, deep dives and the practical difficulties of working under cold, turbid, current-swept and other inhospitable conditions.

Despite the drawbacks, a great deal has been learnt by diving biologists in the last twenty years or so about the ecology of the sublittoral zone and the behaviour and interactions of the organisms living there. This information, linked with data obtained from remote sampling and recording programmes, and from laboratory research, provides an increasingly detailed picture of the subtidal environment of the British Isles and its inhabitants. It is the essence of all these aspects that is conveyed in this book.

2

The Sublittoral Environment

2.1 Coastal and sea-bed topography

The sublittoral zone includes the whole of the continental shelf, from the low tide mark to depths of around 200 m. The subtidal is considered here as synonymous with the shallow sublittoral, which is generally regarded as the area between the low tide mark and around the 50 m depth contour. In some areas, particularly in the west and south-west, this contour is only a few hundred metres from the shore, but more frequently occurs several kilometres away. In the eastern Irish Sea and the central and southern North Sea the sea-bed shelves extremely gently and the 50 m contour is not reached for 100 km or more. The southern North Sea in particular is relatively shallow over a wide area, although there are deeper pockets in places (Fig. 2.1).

Coastal and sea-bed topography are two important factors that determine the biological nature of benthic marine ecosystems. In this context nearshore ecosystems show more complexity than offshore ones because of the way physical features of islands or mainland modify the environment. Many nearshore areas lie along open coastline, but others are slightly, partially or completely enclosed by land and in western Scotland in particular, where the coastline is indented and there are numerous islands, many open-ended channels are formed. Narrow inlets (rias) occur in south-west England and Ireland. Some are small, but one of the largest, at Milford Haven, penetrates more than 20 km inland. A typical ria is steep sided, with strong tidal currents at the entrance, but little diluted by freshwater. Sea lochs are more enclosed than rias, and the outlet to the sea is comparatively small in relation to their overall size. Fast tidal streams sweep through the loch entrance, and conditions in these rapids contrast with the much stiller conditions found at the head of the loch. Incomplete interchange of water between loch and open sea may lead to brackish conditions in the enclosed area.

Coastal lagoons are cut off from the sea by low-lying banks of sand or shingle and water movement tends to be minimal. Lagoonal waters are seldom uniformly saline and there may be areas of reduced or increased salinity away from the entrance to the sea. Estuaries differ from lagoons in that they have a relatively open mouth and a considerable input of freshwater at the landward end. There is typically a wide range of salinities, and often distinct horizontal and vertical salinity gradients. Intermediate between lagoons and estuaries, and combining features of both, are semi-enclosed bays. In these areas accretion and

Fig. 2.1 Sea areas and depth contours around the British Isles. Adapted partly from Lee and Ramster (1981).

erosion of mobile sediments is a continual process, and illustrates the dynamic nature of these parts of the coastline.

2.2 Nature of the sea-bed

The sea-bed around the British Isles is far from uniform; its nature determined by the immediate underlying geological structure, and the amount and type of deposited material. Soft, sedimentary bedrock such as chalk and boulder clay, characteristically found in south-east England, tends to erode rapidly underwater and produce a fairly flat, unstable sea-bed. Thus the striking cliffs at Brighton, Dover and Flamborough Head are not matched underwater but continue as low-lying bedrock or boulder reefs. Limestones and sandstones occur right round the British Isles and are also sedimentary rocks, but are harder and less prone to erosion. However, all are laid down in layers, and the different strata erode at different rates according to their hardness and orientation. In this way reefs, ridges and cliffs are formed underwater and often show secondary erosional features such as caverns, crevices and ledges. Granite, basalt and other hard, igneous rocks from the immediate sea-bed off much of south-west Britain, Scotland and Ireland, and in places give rise to spectacular pinnacles, stacks and steep cliffs underwater. These rocks may erode and split along natural faults, but their hardness prevents them from becoming highly fissured.

In many areas the sea-bed is covered by deposits such as mud, sand, gravel or shell remains, which can cover the underlying bedrock by a few centimetres or many metres. These deposits may occur in localized pockets such as sheltered gullies or bays, or extend over large areas to form sublittoral plains. The gross distribution of mobile deposits around the British Isles is dependent on the action of currents and waves on material deposited over a period of many thousands of years (Fig 2.2).

Mobile deposits continue to be fed into the sea from the land, and a further source of sediments is from the erosion of coastal and marine rocks. For example Devonian slates off the Devon and Cornwall coasts contribute shaly materials to inshore deposits, and these finally break down to silt and clay. In a similar way, there is renewal of sands from the break-up of shells. Calcareous materials are present in nearly all deposits, and often accumulate to form characteristic shell gravels consisting mainly of bivalve shells, such as *Glycymeris*. Coralline gravels may also be formed, for example from the remains of the coralline alga *Lithothamnion calcareum*. Finally, there is a continual input of sediments in the form of organic debris from faecal material and the decay of dead organisms, both in the water column and on the sea-bed.

As a consequence of sorting, mud and fine sand has accumulated on the floor of the North Sea and, locally, has built up in estuaries, lagoons, harbours, deep offshore pockets, and other sheltered places where there is very little water movement (less than about 5 cm sec^{-1}). In contrast, where water movement is strong (i.e. more than 100 cm sec^{-1}) pebbles, shale, shell gravel, sand and other coarse materials are scoured from the sea-bed, leaving exposed bedrock. This has happened in particular off western coastlines. It has also occurred in the central part of the English Channel, where much of the sand originally available for redistribution following the post-glacial rise in sea level has been transported westwards by currents of 160 cm sec^{-1} or more (Holme and Wilson, 1985). In many parts of the English Channel there is only a thin veneer of sediment, about 10–15 cm deep, overlying the rocks. The sea-bed in parts of the Bristol Channel that experience current speeds greater than 3 knots (156 cm sec^{-1}) consists of rocks and boulders that have been worn smooth by the tidal scour. As the current slackens gravel begins to accumulate, especially in hollows in the bedrock, and small gravel waves may be formed (Warwick and Davies, 1977). At current speeds of 1.5–2.5 knots (78–130 cm sec^{-1}) sand

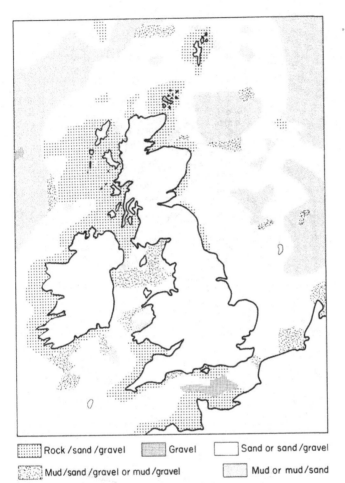

Fig. 2.2 Characteristics of the sea-bed around the British Isles. Adapted from Lee and Ramster (1981).

Rock/sand/gravel   Gravel   Sand or sand/gravel

Mud/sand/gravel or mud/gravel   Mud or mud/sand

ribbons a few centimetres thick occur, while at lesser speeds large sand waves are formed. In areas where currents are weak, fine to muddy sand is deposited.

2.3 Patterns of water movement

2.3.1 Ocean currents

The greatest influence on gross patterns of water flow around the British Isles stems from the Gulf Stream, which originates in the western Atlantic Ocean. The main thrust of this current circulates down to the west coast of Africa, but a northerly offshoot, the North Atlantic current, is deflected upwards towards the Bay of Biscay. It eventually reaches western and northern coastlines of the

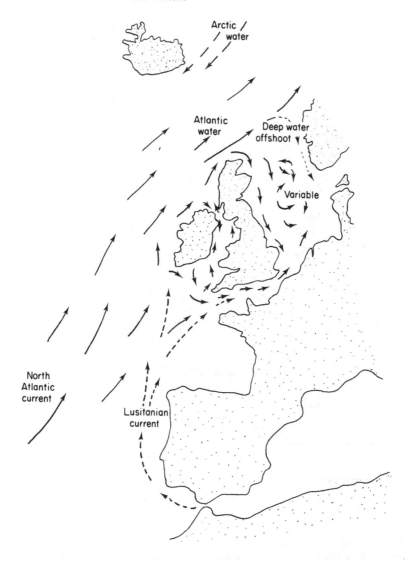

Fig. 2.3   Major patterns of water movement around the British Isles.

British Isles where it is known as the North Atlantic drift (Fig. 2.3). Another source of warm water is the Lusitanian or Mediterranean current. This has a deep outflow from the Mediterranean and gradually moves towards the surface, joining the North Atlantic drift current and finally upwelling off the west coast of the British Isles. There is a steady residual flow of water into the southern North Sea from the English Channel and slight penetration into the northern North Sea by the North Atlantic drift. Currents in the central North Sea are variable and essentially wind-driven, and the main outflow from the North Sea is along the Norwegian coast to the north.

2.3.2 Tidal currents

Coastal waters of the British Isles experience relatively strong tides originating from tidal movements in the North Atlantic Ocean which are magnified as they reach the continental shelf. Particularly strong tidal streams occur where water is funnelled through channels such as the Dover Straits or the entrances to the Irish Sea. Localized tide races flow around headlands or promontories and through narrow gaps between rocks and islands. In these situations flow rate generally reaches 2–3 knots (104–156 cm sec^{-1}) and may be as high as 5 knots (260 cm sec^{-1}). In the Bristol Channel, where the water is pushed forward into an ever narrowing space, tidal streams approaching 7 knots (364 cm sec^{-1}) are experienced. Smaller inlets on western and south-western coasts experience similar small scale surges, and tidal streams up to 8 knots (416 cm sec^{-1}) may occur in the entrance to sea lochs. In most of the North Sea tidal streams up to 3 knots (156 cm sec^{-1}) are possible, but in general they do not exceed 1 knot (52 cm sec^{-1}).

Superimposed on this overall picture are variations produced at a local level which are not indicated on bathymetric charts. Tidal currents may be reduced, deflected, accentuated or in some other way altered by small scale topographical features of the sea-bed, and these changes have an impact on sublittoral communities.

2.3.3 Waves

Waves are generated by winds blowing across the surface of the water and their size depends on wind speed, length of time the wind prevails, and the distance over open water the wind has blown (fetch). Coastal areas facing west are exposed to the open Atlantic and, in the case of prolonged storms, could be subjected to waves with a height of up to 35 m. Maximum wave height off the central east coast does not exceed 20 m, while in the south-east it remains below 10 m, even under severe conditions (Lee and Ramster, 1981). Average wave height is, of course, less extreme, but the general pattern still remains, though modified by topographical features of the coastline and the sea-bed.

Surface waves travelling over deeper areas of the continental shelf have no direct physical impact on the sea-bed. This is because the orbital motion created on the surface diminishes logarithmically with depth and has died away before it reaches the bottom. The situation is different in the coastal fringe, where the sea-bed may be consistently or periodically affected by wave-induced water movement. Waves are felt to a depth approximately equal to their wave length, but their motion is modified by both the presence and slope of the sea-bed.

Another characteristic of waves is that as they approach the shore they are deflected towards it as a result of refraction. One outcome of this is that their impact is spread disproportionately along the coastline, according to coastal features. The energy of the wave front becomes concentrated on headlands and promontories whilst it is correspondingly diminished in bays. In coastal areas where the sea-bed is rocky and uneven there may be further modifications in the

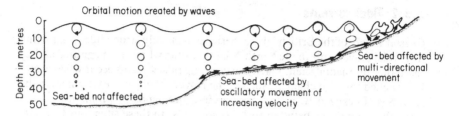

Fig. 2.4   Simplified illustration of the changing effect of surface waves of moderate strength on a sloping sea-bed. Modified from Riedl (1964), and Hiscock (1983).

degree of exposure to wave action. In this way reefs and pinnacles of rock can cause very localized variations in water movement from the lee side to the exposed side.

2.4 Temperature

The temperature of coastal waters around the British Isles varies according to geographic location, season and depth. The south-west is consistently warmer throughout the year while the central east coast experiences not only the coldest temperatures but also the widest fluctuations. This pattern is determined by the overall gradient of ambient temperatures from north to south, the warm water currents flowing along western coastlines and the influence of the continental land mass to the east. It is modified on a local scale by factors such as inflow of river water and patterns of water movement (Fig. 2.5).

Fig. 2.5   Variations in mean surface temperature (°C, winter–summer) around the British Isles. Based on information in Lee and Ramster (1981).

Vertical layering (stratification) may occur in certain areas, but the extent to which this happens depends on the amount of heat falling on the surface of the sea and the degree of mixing which subsequently takes place. Warm water is less dense and remains at the top of the water column unless disturbed and mixed by water movements. In winter there is little or no temperature stratification in the shallow sublittoral, and in spring and summer the temperature of surface waters in many areas is only a degree or two higher than it is at 40–50 metres. However, in areas where there is relatively little mixing, a well-defined interface (thermocline) develops in the spring and summer months. A sudden drop of several degrees is experienced over just a few metres, dividing the water column into a warm upper zone, and an underlying layer not directly affected by the sun's rays.

Horizontal discontinuities have also been discovered in recent years, although the temperature difference between the two adjacent bodies of water is much less marked than it is at the thermocline. However, a temperature change of 1 °C per kilometre may occur across the boundary zone known as the 'front'. These fronts tend to form roughly in the same areas each year, and are particularly evident where oceanic waters meet coastal waters (Fig. 2.6). For example one appears in the western Irish Sea during the summer, marking the boundary between stratified offshore waters and mixed inshore waters.

2.5 Light

The behaviour of light when it reaches the surface of the sea, and again as it

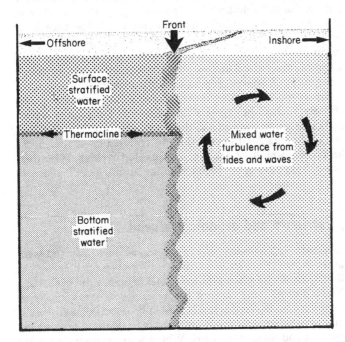

Fig. 2.6 Diagrammatic representation of a frontal system.

passes through the water column, is a complicated one. In the British Isles the intensity of incident light varies considerably with season, latitude and time of day, and is also affected by cloud and atmospheric conditions. Some or all of the light falling on the surface of the water is reflected away, unless the sun is directly overhead and the sea completely calm. Light that penetrates the water is then absorbed or scattered, but the rate at which this happens depends on turbidity and the type of particulate matter present. In westerly areas the sea-bed at 50 m may be clearly illuminated while in the North Sea, where turbidity is consistently high, it may, under optimal conditions, be only dimly lit at 10 m. Locally, a plankton bloom in surface waters or an inflow of silty river water can act as an effective barrier and prevent light reaching the sea-bed. Illumination is often significantly reduced in rocky areas where cliffs and overhangs cause shading, but in sandy areas illumination may be enhanced by reflection from the sea-bed. The quality of light is also affected because different wavelengths are absorbed at different rates. Coastal waters around the British Isles contain green and yellow pigments from phytoplankton and the products of organic decay, and blue light is selectively absorbed, resulting in a general green cast. The colour of the sea becomes bluer as turbidity decreases and the water becomes cleaner, a condition seldom seen at the eastern end of the English Channel or in the North Sea.

2.6 Salinity

Western approaches are influenced by water originating from the Atlantic Ocean and salinity is normally around 35 o/oo (parts per thousand), but in much of the North Sea the level is held down to around 34 o/oo because of dilution by river water. Even lower levels are found in the eastern Irish Sea. Marginal changes in this overall pattern occur between winter and summer and between surface and bottom waters, but the most marked differences are found in enclosed or semi-enclosed waters. Freshwater inflow depresses the level of dissolved salts, while evaporation causes it to rise, and salinity is determined by these and other factors, such as the degree of tidal flushing. The salinity regime in estuaries is particularly complicated and there are horizontal as well as vertical gradients which fluctuate with every tide. In enclosed or semi-enclosed coastal areas with a freshwater input isoclines may develop in parallel with thermoclines, marking a discontinuity between less saline surface water and denser bottom water.

2.7 Gases, nutrients and organic material

Atmospheric gases enter into solution across the sea–air interface, and well-mixed surface layers above the thermocline have a uniformly high dissolved gas content. In well-lit areas where photosynthesis is taking place the percentage saturation of dissolved oxygen may exceed 100%. In deeper areas, and especially below the euphotic zone, oxygen levels tend to be lower because there is no direct input from photosynthetic processes, yet oxygen is being used during respiration of bacteria and other organisms. Where water becomes stratified during summer an oxycline may develop, where a rapid drop in oxygen content is apparent. This condition may arise in sheltered sea lochs but is less likely in the open sea.

Even more dramatic depletion occurs in sediments, especially in fine mud with a high organic content where there is considerable bacterial activity. Here anoxic conditions can develop within a few centimetres of the water-sediment interface, but this condition is likely to occur only in sheltered locations. Gaseous carbon dioxide and its dissociation products are present in inshore waters at consistently high levels. Carbon dioxide is essential for plant growth but is unlikely to be a limiting factor in the sea. Carbonates are incorporated as skeletal material by a wide range of marine animals and some plants.

Nitrogen and nitrogenous compounds are required for the growth of marine organisms, as are various salts, trace elements, dissolved organic substances and biologically active compounds. Coastal waters are generally well supplied with these essentials, even though there may be seasonal and regional fluctuations. Levels of around 10–50 mg per litre of organic matter are typical of coastal waters around the British Isles. Some of this is held in suspension as particulate organic matter (POM), which is derived from the breakdown of faecal and decaying material, and also contains some inorganic substances. The rest is finer material referred to as dissolved organic matter (DOM), which includes various carbohydrates and amino acids.

3

Environmental Factors Affecting
Species Distribution

3.1 Introduction

The distribution of marine species locally and on a wider, geographic scale, is governed by various environmental and biological factors. In some cases a single factor determines whether or not a species will occur in a certain area or habitat, but often the situation is not so clear cut. For benthic species the topography of the sea-bed and type of substratum are of paramount importance in influencing distribution, but habitats that look superficially alike in this respect may support a completely different mix of species. It is quite possible to find dissimilar benthic communities on, for example, two apparently similar bedrock habitats within a few hundred metres of each other. In a case such as this the differences cannot be attributed to geographical location, but are due to one, several or a combination of local factors. Illumination, depth, turbidity, sedimentation and water movement all play their part, and species distribution is also influenced by the biota itself. Differences in the distribution of pelagic species is not so marked because environmental gradients in open waters are generally less pronounced.

This chapter is concerned primarily with the physiochemical environment, and the ways in which environmental factors interact to affect the distribution of species. More detailed analyses of the biological forces at work at community level are described in following chapters.

3.2 Biogeographical aspects

The distribution of species in relation to geographical location is a subject that has generated many classical studies, and the broad affinities of our marine flora and fauna have been known for many years. Only recently, however, have some of the more detailed facts emerged. It is generally accepted that, with regard to surface waters (0–200 m depth) temperature is a critical factor in determining the geographic spread of species. Three major areas are recognized: warm waters (temperatures consistently above 18–20°C), cold waters (temperatures consistently below 5°C) and temperate waters (temperatures between 5–18°C, fluctuating seasonally). Further subdivisions have been made, again based primarily on temperature regimes, but also related to barriers or pathways created by water currents, land masses or island chains.

The British Isles lie within the cold-temperate (Boreal) province but are overlapped by the warm-temperate (Lusitanian) province in the south-west and the cold (Arctic) province in the north (Briggs, 1974). This is one of the reasons why, although the majority of sublittoral species have a cosmopolitan distribution, others clearly do not (Fig. 3.1). When comparisons are made between similar sublittoral habitats around the British Isles the main differences that emerge are between southern and western areas on the one hand and northern and eastern areas on the other. A feature of the south-west is the high proportion of warm-water species which, because of the Lusitanian current, have been able to penetrate from the Mediterranean region. A few benthic species appear periodically in western approaches but fail to maintain a permanent population in the British Isles, others have a precarious hold in the extreme south-west, while the rest penetrate southern and western coastlines to a varying degree. Pelagic species show similar distributional patterns, as illustrated by the regular appearance of siphonophores, pelagic tunicates and other exotic species such as the Wreckfish, *Polyprion americanus* and trigger-fish, *Balistes carolinensis*. Conversely there are species such as the feather-star *Antedon petasus*, the starfish *Henricia sanguinolenta* and the Norway Haddock, *Sebastes viviparus*, which are cold-water species restricted to the northern regions of the British Isles.

3.3 Substratum type

It is not surprising to find that benthic species, which live in or on the sea-bed, show distinct distributional patterns in relation to the type of substratum. The distribution of many of the semi-pelagic species, which swim in open water, but visit the sea-bed to feed, bears a similar relationship. Broadly speaking the substratum falls into two categories; hard (non-particulate) and soft (particulate), but it can then be subdivided into several major types. Further divisions can go

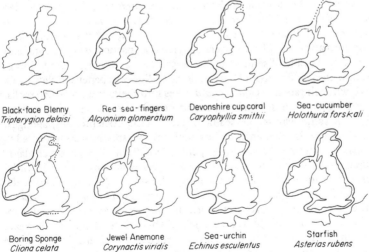

Black-face Blenny
Tripterygion delaisi

Red sea-fingers
Alcyonium glomeratum

Devonshire cup coral
Caryophyllia smithii

Sea-cucumber
Holothuria forskali

Boring Sponge
Cliona celata

Jewel Anemone
Corynactis viridis

Sea-urchin
Echinus esculentus

Starfish
Asterias rubens

Fig. 3.1 The geographical distribution of some benthic organisms around the British Isles.

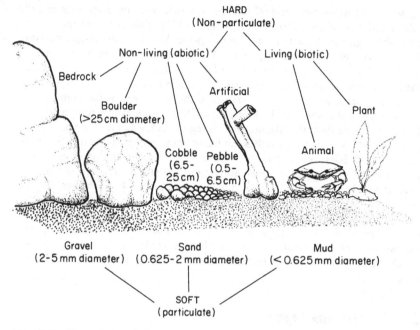

Fig. 3.2   The major types of benthic substrata.

on almost endlessly, depending on the scope of the study and the need for descriptive detail (Fig. 3.2).

The degree of specificity shown by different organisms for particular substrata depends on the mode of life and requirements of the species involved. Animals such as the sea-potato *Echinocardium cordatum*, the holothurian *Neopentadactyla mixta*, the Norway Lobster *Nephrops norvegicus* the Seamouse *Aphrodita aculeata*, the razor shell *Ensis siliqua*, and many other polychaetes and bivalves are all burrowers restricted to particulate substrata, and often to a certain grade of sediment. Similarly, boring organisms are found in association with specific types of hard substrata. Piddocks penetrate clay, chalk and wood, while the sponge *Cliona celata* bores into shells and limestone rocks. Sessile and sedentary organisms are found only in areas where they can fix themselves in some way to the sea-bed. Most algae, coelenterates, sponges, bryozoans and ascidians are anchored by stalk or base, and thus are restricted to hard substrata such as rocks, boulders, pebbles and the surfaces of other plants and animals. A few species, for example the mussels *Mytilus edulis* and *Modiolus modiolus*, the ascidian *Molgula manhattensis*, the sponge *Suberites domuncula*, the bryozoan *Pentapora foliacea* and serpulid worms such as *Filograna implexa*, all of which are normally attached to hard surfaces, may extend on to soft substrata because they are able to live unattached where water movement is weak. However, the early life of these organisms depends on the attachment of the post-larval stage to a shell or fragment of stone, which is subsequently outgrown. Animals such as these are important because they form miniature

living reefs which can then be colonized by other species requiring hard surfaces for attachment (p. 51). In contrast there are sedentary organisms
including the sea-pen *Virgularia mirabilis*, the anemone *Cerianthus lloydii*, and
polychaetes such as *Sabella penicillus* and *Lanice conchilega* which are specifically adapted to soft substrata and have the tube or basal part of the body
acting as an anchor and thrust deep into the sediment.

Some of the animals that creep or crawl over the sea-bed, such as the spider
crab *Maja squinado*, the brittle-star *Ophiothrix fragilis* and the starfish *Asterias
rubens* are found on a wide range of substrata, but others are more restricted.
Often this is because of the need for shelter from predators, water movement or
other environmental stress. For example the sea-cucumber *Pawsonia saxicola*
lives in rocky crevices, the Tompot Blenny *Parablennius gattorugine* and the
squat lobster *Galathea squamifera* require similar hiding places, while the
lobster *Homarus vulgaris*, although it can excavate burrows in soft substrata,
will do so only if there is protection from rocks or boulders. Conversely the
Masked Crab *Corystes cassivelaunus* lives only in sandy areas where it is able to
dig its way under the surface for protection.

Finally, there are a large number of mobile surface dwellers which are associated with certain substrata because of the food they eat. Thus most nudibranchs, which feed on attached organisms such as hydroids and sea-squirts,
are found in areas where the sea-bed is hard, as is the sea-urchin *Echinus esculentus*, which grazes on sessile organisms. In contrast the necklace shell *Natica
alderi* and the whelk *Buccinum undatum* are found in sandy areas where they
prey on polychaete worms and bivalve mussels.

Species assemblages make up communities, and so it is easy to see why, with
many species having distinct habitat preferences, different types of substratum
support distinctly different communities. Sometimes there is a wide variability
in the nature of the sea-bed over a small area, leading to a mosaic of different
benthic communities which can be seen with a single sweep of the eye.

3.4 Light and depth

Light plays a key role in determining the distribution of sublittoral species,
operating chiefly through its effect on primary production and the growth of
algae. Algae live only within the upper layer of the sea known as the photic
zone, where there is sufficient light for photosynthesis to occur. Within this
zone there is a critical depth which marks the point at which algae can obtain
sufficient light energy to grow and reproduce; in other words where, over the
life span of the alga concerned, energy loss via respiration is balanced by energy
gain through photosynthesis. This absolute limit for growth differs from the
compensation depth which is the depth, at any particular point in time, at
which net production by individual cells is zero. The critical depth for kelps
and phytoplankton corresponds roughly to the depth at which irradiance levels,
averaged over the whole year, fall to about 1% of their value at the surface.
Small seaweeds, especially crustose red algae, are successful at lower light
intensities, so the critical point for these types is deeper, corresponding to irradiance levels of around 0.05% of surface values. Under favourable conditions
the critical depth for benthic algae may reach 40–50 m in some waters around

Isles, but in other areas it is considerably less. Within this zone, and providing that other factors are not limiting, then algae will occur in various situations. Phytoplankton are adapted for life in the water column, while the sea-bed is colonized by micro- and macro-algae. Benthic micro-algae such as diatoms and flagellates are widespread in their distribution, but probably play a particularly significant role in sediments because apart from the plants that have become detached and are drifting loose on the sea-bed, they are the only source of primary production in these areas. At present, however, little is known about the precise effects these micro-algae have on the distribution and abundance of sediment faunas. In contrast, the part played by macro-algae in shaping sublittoral communities is more readily studied and better understood.

In well-lit areas where the sea-bed is solid and stable, and other factors are favourable, an algal-dominated community develops. In recent terminology this is referred to as the infralittoral zone. Animals are never totally excluded from the infralittoral but may be restricted to the darker recesses, or to the surface of the algae themselves. As light levels fall with depth, shading or turbidity, or perhaps a combination of all three, cover drops and animals predominate in what is called the circalittoral zone (Fig. 3.3). There are qualitative as well as quantitative changes in algal distribution in response to changes in

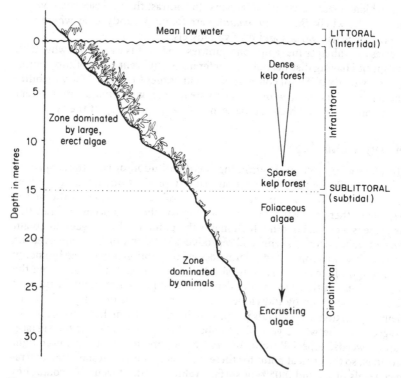

Fig. 3.3 Zonation patterns at a moderately exposed rocky subtidal site. After Hiscock and Mitchell (1980).

illumination because different types of algae have specific requirements and utilize light of different wavelengths. Green algae are restricted to shallow water and are least significant in terms of cover and species diversity. Brown and red algae colonize a much broader band, the vertical width of which is determined by the species involved and the level and quality of illumination at that particular locality. Red algae make use of green light and thus are able to penetrate to the greatest depths.

Kelp forests extend to a depth of about 16 m in clear, west coast localities, 10 m off the more turbid coast of north-east Scotland, but rarely ever form forests at the eastern end of the English Channel. Individual plants are found to about 30 m, 18 m and 9 m respectively in the same areas. Other algae can grow under lower light regimes, which enables them not only to colonize the shaded substratum beneath the kelp canopy, but also continue beyond the lower limits of kelp growth.

Light plays a critical although indirect role in determining faunal distribution through its control on the growth of algae. Where algae occur then herbivores and grazers follow, and so on, up the food chain. Algae also provide cover for various mobile animals, ranging from crustaceans to molluscs and fish, and suitable surfaces for attachment of sessile organisms. The diverse fauna and flora associated with kelp plants is a classic example (p. 72), but other algae, provided they are not too small and fragile, are also colonized, especially by small hydroids and bryozoans. Conversely, the presence of algae means that the abundance and diversity of sessile animals may be reduced as a result of competition.

Light also directly affects faunal distribution by modifying and controlling the physiological and behavioural processes of the animals themselves. Many animals are known to orientate themselves to light, particularly to intensity gradients, but also to direction, wavelength and polarization. The way that individual animals react to light is complex because their responses may change according to age, physiological and nutritional state, or some other endogenous or environmental factor. Without careful experimentation it is generally impossible to disentangle the precise part played by light from other interrelated factors that determine the distribution of animals. For instance, the anemones *Corynactis viridis* and *Actinia fragicea* are both typical of shaded habitats, but it is not known if this distribution is due to intolerance of bright light, or whether it is governed primarily by other factor(s) such as the amount of siltation, water movement or competition.

3.5 Water movement

Water movement created by tidal currents and wave action has a considerable influence on the distribution of marine organisms, operating at individual and community levels in a number of ways. In the pelagic zone plankton is moved bodily from place to place by water currents, and this affects the distributional patterns of both permanent and temporary plankton species. The movement of the temporary plankton species, which includes the larvae and spores of numerous benthic organisms, clearly must affect the ultimate distribution of these species on the sea-bed.

Water movement created by waves and currents also has an important influence on the benthos; affecting the physical surroundings such as sediment grade, turbidity and amount of deposited silt as well as the organisms themselves through physical stress (Fig. 3.4).

It is important to emphasize at the outset that water movement is one of the more variable of the environmental parameters and that, while it is possible to consider 'average' exposure, it is also essential to recognize that extreme conditions have profound effects, even though they may persist for a relatively short length of time. In this respect, water movement due to wave action is the more erratic because it fluctuates considerably on a seasonal basis. Movement caused by tidal currents waxes and wanes according to a regular lunar cycle. Furthermore it is not only the strength but also the type of movement that affects the distribution of marine organisms. Uni-directional, multi-directional and oscillatory movements each presents a different type of stress or confers particular advantages.

In extremely sheltered habitats where water movements are very weak (less than 10 cm sec^{-1}), benthic organisms can settle, grow and move freely without being dislodged or damaged by swell, surge and currents, or scoured by swirling sand or stones. On the negative side, however, such conditions reduce mixing, thereby minimizing renewal of dissolved gases, nutrients and other water-borne essentials, especially planktonic food. In extreme cases, for example in enclosed bodies of water such as sea lochs, where water movement is often less than 5 cm sec^{-1}, lack of mixing can lead to stratification in the water column, development of an oxycline, and oxygen starvation of fauna below it.

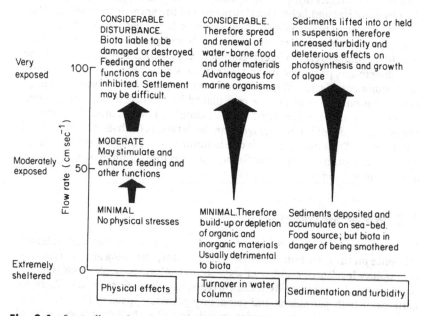

Fig. 3.4 Some effects of varying degrees of water movement on benthic biota (based on information in Reidl, 1971 and Hiscock, 1981, 1983).

Also, waste products are not swept away and eggs, sperm or larvae are not lifted up and dispersed. Instead, silt and other fine sediments, if present in the area, will settle and accumulate. For surface-dwelling deposit feeders such as holo-thurians and terebellid worms this deposited material, provided it is organically rich, is an important source of food. But it is a problem for most benthic orga-nisms because it tends to clog and overload the delicate structures associated with feeding and removal of debris. It can also smother respiratory and photo-synthestic surfaces so that the efficiency with which they operate is cut. Furthermore, the accumulation of silt on hard surfaces creates a barrier which larval forms find difficult to penetrate. Because of these problems the biota of extremely sheltered areas is restricted in terms of species diversity.

Moving water produces opposite effects and moderate flow or surge (10–100 cm sec[-1]) is generally advantageous both to open water and benthic species. Mixing and renewal of inorganic and organic material is ensured, waste products are prevented from accumulating, reproductive stages are dispersed and various physiological processes may be enhanced. For, example, the flow of water through the body of sponges and ascidians may be boosted by the presence of external currents (Vogel, 1974), thereby increasing efficiency of feeding, respiration and excretion. Moderate water movement lifts fine sedi-ments from the sea-bed, thus ensuring that potential food is kept in suspension and that the biota is not smothered. A negative element here is that suspended sediments increase turbidity and therefore inhibit photosynthesis and affect the distribution of algae. Taken overall, however, there are more beneficial than harmful effects, and areas subjected to moderate water movements generally support a rich and diverse benthic flora and fauna.

Beyond a certain point, water movements bring more disadvantages than advantages and species diversity begins to fall. Some species will be adversely affected before others because different species have markedly different tole-rances. For example the Snakelocks Anemone *Anemonia viridis* was found to retract its tentacles at uni-directional flow rates of around 20 cm sec[-1] and had the base dragged from the substratum at flows in excess of 40 cm sec[-1]. In contrast the Plumose Anemone, *Metridium senile*, which is adapted to fast flow rates, did not retract the tentacles until water movement exceeded 75 cm sec[-1], and showed no signs of becoming detached (Hiscock, 1983). In the same series of experiments the unattached brittle-star *Ophiothrix fagilis* was swept off the substratum when the flow rate approached about 30 cm sec[-1]. Mobile species also respond to water flow, and station themselves in the most advantageous position. For example young Two-spot Gobies, *Gobiusculus flavescens*, which are typically found amongst kelp and rock, gather in regions that provide suffi-cient shelter to prevent them being washed away, yet enough water movement to bring in a continuous supply of plankton (Potts and McGuigan, 1986).

When water particles move over the sea-bed at speeds in excess of about 100 cm sec[-1] the benthic flora and fauna are undoubtedly exposed to considerable physical stress. Animals and plants are liable to be lacerated and abraded by sand, gravel, pebbles and even small rocks scoured from the sea-bed, and to be damaged as they are dragged against each other or against the sea-bed. In extreme conditions kelp plants, sponges, hydroids and other organisms may be partially or completely ripped from hard substrata, and animals such as worms, crabs, bivalves and heart-urchins washed out from upturned sediments. Some of these mobile sediment-dwellers are protected by shells and, if they are not too damaged by the buffeting they receive, have a chance of re-establishing themselves when conditions become calmer. Rees *et al.* (1977) described redistribution of adult animals after storms, and pointed out that violent water movements may play a part in the formation of benthic communities as well as their destruction. Several anemones and sponges can also regrow from remnants left on the rocks after storm damage.

3.6 Salinity and water quality

Water quality is a useful descriptive term, but one which unfortunately has a number of different and rather imprecise meanings. In the sense in which it is used here it relates to the overall environmental characteristics of a particular body of water. In other words, factors such as turbidity, levels of organic matter, pollutants and dissolved gases are considered together, for their combined (synergistic) effect, rather than in isolation.

Although it is difficult to categorize sea water into a particular 'grade', a biological definition of water of low quality is that which puts marine species under stress of some sort. Enclosed, semi-enclosed or sheltered areas are particularly prone to suffer from water of poor quality because of lack of circulation and renewal. As a general rule a drop in water quality or salinity leads to changes in species composition as well as a reduction in the number of species and, at the lowest end of the scale, a reduction in overall biomass. This type of effect can be seen at a local level in places such as Liverpool Bay and Southampton Water, where water quality is reduced mostly as a result of accumulated domestic and industrial discharges. At the eastern end of the Bristol Channel, where turbidity is high, layers of fluid mud a few metres thick form near the sea-bed during each neap-tide. These transient layers formed beneath the water column with its normal load of suspended material result in an impoverished soft-bottom community (Warwick, 1984). On a wider geographic scale, water quality at the eastern end of the English Channel and in the North Sea is generally considered to be of lower quality than western areas, mostly because of its higher turbidity and lower salinity. Sheppard *et al.* (1980), in a study of the fauna inhabiting holdfasts of the kelp *Laminaria hyperborea*, found gradients in total numbers of species and species diversity which correlated closely with environmental gradients such as turbidity and heavy metal pollution.

3.7 A community approach

Marine biologists have put considerable effort into community analysis and classification ever since the classic work by Petersen (1913), who grouped benthic faunas on the basis of common species which could be used as indicators. In this way it proved possible, by looking for these indicator species in only relatively small samples, to predict the type of community from which the sample was taken. This basic framework is still in use today, although has been refined and expanded in the light of more recent work, and computer analysis of the data has enabled cluster patterns to be seen (e.g. Warwick and Davis, 1977). The scheme is used especially for soft bottom communities, partly because sampling in this habitat has been in progress for longer and use of dredges, grabs and core samples has provided many data. For example, in the North Sea, classification analysis of all benthic invertebrates trawled from 48 stations revealed, in addition to a basic division between the northern and southern North Sea benthos, three benthic regions in the southern North Sea, and four in the northern North Sea (Dyer *et al.*, 1983).

In addition to this faunistic approach the 'community concept' has been looked at from different angles, and in particular from an environmental viewpoint. For example, Jones (1950) used bottom type as the key element, while Glemarec (1973) classified communities primarily according to the temperature regime. Communities can also be described and classified in terms of the way they function, regardless of species composition. Erwin (1983) has taken a step further in trying to disentangle the problem by proposing a hierarchical system of classifying biological assemblages rather along the lines of that erected by Linnaeus to classify living organisms. Thus, for example, an assemblage could be classified as follows: euhaline; coastal; bedrock; vertical face community; *Pachymatisma johnstonia* facies. The facies is descriptive of a single 'indicator' species in this case, but in other instances an association of characteristic species might be used to describe the community. No doubt in the years to come this and other systems will be refined and more ideas will emerge. However, there are many who believe it is impossible to try and draw rigid boundaries around 'discrete' communities because each of the species present will have a different distribution pattern, leading to overlapping and gradation between one 'community' and another.

In this book 'community' is used in a broad sense to describe the assemblage of plants, animals and other organisms which inhabit a particular area. At one end of the scale it is descriptive of the biota of whole habitats, at the other end it focuses on discrete habitats such as kelp holdfasts, or microhabitats such as the crevices in a rock face or the spaces between sand grains.

4

Benthic Communities: Hard Substrata

4.1 Introduction

Hard substrata are united by their non-particulate nature but are enormously variable as far as size, shape, profile, texture, hardness and stability are concerned. At one end of the size scale are small boulders, pebbles and shells, at the other end, bedrock outcrops, massive boulders, pier piles and wrecks. Profiles range from flat on the one hand to vertical or underhanging on the other, and may be uniform, but more often are complicated by the presence of caverns, gullies and holes. Some surfaces are smooth, others rough or deeply fissured, and the substance beneath may be hard and impenetrable (granite, concrete), or relatively soft (chalk, wood). The stability of a non-particulate substratum, in terms of its tendency to move, fragment or disintegrate, is determined by its form and structure, and life history in the case of biotic substrata, and influenced by external conditions, particularly water movement.

The algal and animal communities reflect these variations of the substratum, and in addition, are influenced by other environmental factors such as light, depth, turbidity and siltation and by biological interactions within the community itself. Some broad dividing lines can be drawn between different types of non-particulate substrata, and the communities that they support but, with so many variables involved, it is not surprising that categorization into 'typical' communities is made difficult.

In this chapter distinctions are drawn between immobile, mobile, artificial and living substrata. Bedrock and boulders are immobile, as are cobbles and pebbles where water movement is minimal. This permanent stability of the substratum influences strongly the type of community present because it encourages biotic continuity and gives the potential for a mature, climax community to develop (Chapter 6). However, whether this potential is realized depends on other environmental and biological factors. Wrecks and other artificial substrata are also described separately, because the associated communities often show certain distinct features that set them aside from neighbouring hard-bottom areas.

The biota of sublittoral hard-bottoms can be sampled with trawls, dredges and grabs, but these rather crude methods have their limitations, and much more has been learnt about the communities of the shallow rocky sea-bed since the advent of scuba diving. In particular, *in situ* studies are providing detailed information on the behaviour of organisms and their interrelationships; two aspects which are difficult to investigate using traditional techniques.

4.2 Immobile soft rock

Communities associated with chalk, clay and similarly soft rocks have several unique features, the most obvious of which is the prevalence of infaunal animals which take up residence within, rather than on the surface of the substratum. Some of the softer limestones also support an infaunal community, although the emphasis is on epifauna.

Prominent amongst the infauna are a number of species of bivalve mollusc which are adapted specifically for burrowing, and are often present in huge numbers. *Hiatella arctica* and piddocks such as *Pholas dactylus* and *Barnea* spp. are well known for this habit (Fig. 4.1). The post-larvae settle onto the sea-bed from the plankton and then penetrate the rocks by mechanical means, grinding and etching their way inwards by rotational movements of the shell valves. The distance to which they penetrate is governed by their need to remain in contact with the outside, since they rely on water-borne material for food. This connection is made by elongate siphons, the openings of which lie flush with the surface of the rock. On a different size scale is the polychaete worm *Polydora ciliata* whose tiny U-shaped tubes often riddle the upper crust of soft rocks. The sponge *Cliona celata* is also able to bore into rocks, using a combination of chemical and mechanical means. Immature stages of the boring sponge are largely hidden within the rock itself, with only the inhalent apertures visible as small papillae at the surface. As the sponge grows a large external mass develops, although at the eastern end of the English Channel this species never develops beyond the boring stage.

Holes and tunnels created by boring animals provide shelter for other

Fig. 4.1 Siphons of the piddock, *Pholas dactylus*, protruding from chalk bedrock covered by a thin layer of debris.

animals, particularly anemones, polychaetes, squat lobsters and decapods such as pea crabs and the Hairy Crab *Pilumnus hirtellus*. Some of these crustaceans live in holes that have been bored and then vacated, but others, particularly pea crabs, take up residence in piddock holes still occupied by the piddock itself, and probably maintain this close association throughout their lives. These microhabitats also provide temporary shelter holes for small fish such as the Tompot Blenny *Parablennius gattorugine* and the Leopard-spotted Goby, *Thorogobius ephippiatus*.

The activities of boring organisms can cause or accelerate erosion and collapse of a substratum which in itself, is more prone to erosion than other hard substrata. Physical, rather than biological erosion also takes place, resulting in the gradual removal of surface layers, and under-cutting of boulders or outcrops. This is caused by the scouring effect of the water itself, together with sand, pebbles and other water-borne materials. Clay and chalk often collapse as a result, but some of the sandstones can withstand a high degree of under-cutting. This leads to the formation of small caverns, which provide shelter for larger organisms, such as decapod crustaceans.

The extent to which the sessile epifaunal element of soft rock biota is developed, and the type of organism present is influenced considerably by the erosional properties of the rock surface. This in turn depends on the geological nature of the rock, the degree of exposure to water movement and scouring, and the amount of boring which is taking place. Chalk and boulder clay, which are the softest rocks, may remain virtually uncolonized, except by a transient flora and fauna which is able to develop only when conditions are calm. Foliaceous and filamentous algae grow in relatively shallow, well-lit areas, while a hydroid or bryozoan turf may develop in deeper or shaded locations, and all the species involved tend to have low-drag characteristics. Under more stable conditions a relatively rich biota can develop, and this in itself can lead to a strengthening and consolidation of the surface crust as a matrix of animal and plant material is deposited. This happens on some of the chalk reefs and outcrops in south-east England, although even under stable conditions where there is potential for epifaunal communities to develop as they would on harder substrata, these communities retain a special identity dictated by their geographic situation. Kelp forests, so typical of shallow rocky areas around much of the coastline of the British Isles (section 4.3), are less dense and thin out rapidly with increasing depth. *Laminaria saccharina*, rather than *L. hyperborea* or *L. digitata*, tends to be the dominant species, and an animal conspicuous for its absence is the sea-urchin *Echinus esculentus*. The Jewel Anemone, *Corynactis viridis*, is also missing, while bryozoans, phoronids and ascidians, especially *Molgula manhattensis* are common.

4.3 Immobile hard rock

Some of the limestones and sandstones that occur around the coast of the British Isles, although not as soft as chalk and boulder clay, are soft enough to support a population of infaunal animals. However, the increased surface stability of these rocks means that the epifaunal element will be relatively much better developed than it is on chalk or clay substrata. An interesting exception is the quartz-mica rock found in south Devon, which readily fragments along natural cleavage lines, so disrupting the development of epibiotic communities (Rubin, 1980). Igneous rocks are so hard as to be impenetrable to burrowing animals, and although 'nestlers' are found in cracks and crevices, a true infaunal community is absent. Immobile rocky substrata include, in addition to bedrock and large boulders, certain areas of sea-bed covered with smaller boulders or even cobbles and pebbles where water movement is insufficient to cause these stones to move.

These hard immobile substrata occur widely around the coastline of the British Isles, but the associated biotic communities are far from uniform and reflect differences in geographical location and environmental conditions. There are, for example distinct differences with depth, light penetration and profile. In addition, water movement from tidal currents and wave action plays a major role in shaping community structure, and may be the most important factor. The distribution and abundance of many animal species on rocky substrata around the coastline of Lundy, an island in the Bristol Channel, was clearly correlated with exposure to wave action and/or to tidal streams (Hiscock, 1983). However, the limits of tolerance vary considerably with different species, some showing a clear preference for a particular current or wave regime, others being more widely spread. Another complication in categorizing sites according to exposure is that it assumes an 'average' exposure for any particular site. In reality this is seldom the case, especially with regard to exposed sites where there are nearly always shelter spots in crevices, caverns or amongst the biota itself. Thus a very exposed site can often support discrete communities of organisms intolerant of extreme water movement which live permanently in the shelter spots or retire to them when conditions become unfavourable. One statement that can be fairly safely made is that extremely exposed sites on the one hand, and extremely sheltered ones on the other, both cause stress and result in a restricted biota of tolerant species.

4.3.1 Sheltered sites

Some sheltered rocky sites are found in deep water offshore, but often bedrock in these areas is covered completely with layers of sediment, unless it is steeply sloping. It is more common to find sheltered rocky sites around the margins of lochs and semi-enclosed bays, or in more open coastal situations protected from tidal currents and wave action by irregularities of the sea-bed. Considerable siltation occurs in these areas, particularly where flow rates are 5 cm sec^{-1} or less, and is one of the major factors determining community structure and species composition. Upward-facing surfaces are particularly affected and generally support a sparser community, and a smaller variety of species. In sea lochs sessile organisms may be restricted entirely to vertical and overhanging rocks. Many of the attached animals that occur in sheltered situations have an erect growth form which holds them above the layer of silt, and they also have mechanisms for removing sediment from their surfaces. Other animals take advantage of this. For example, the feather-star *Leptometra celtica*, known only

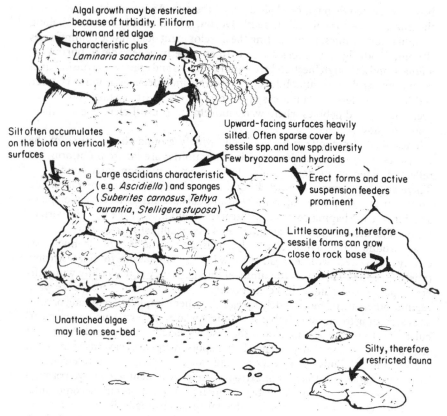

Algal growth may be restricted because of turbidity. Filiform brown and red algae characteristic plus - *Laminaria saccharina*

Silt often accumulates on the biota on vertical surfaces

Large ascidians characteristic (e.g. *Ascidiella*) and sponges (*Suberites carnosus, Tethya aurantia, Stelligera stuposa*)

Unattached algae may lie on sea-bed

Upward-facing surfaces heavily silted. Often sparse cover by sessile spp. and low spp. diversity Few bryozoans and hydroids

Erect forms and active suspension feeders prominent

Little scouring, therefore sessile forms can grow close to rock base

Silty, therefore restricted fauna

Fig. 4.2   Characteristic features of a sheltered rocky site.

from sea lochs and other sheltered situations in western Scotland, may climb on to the branches of sea-fans such as *Swiftia pallida*.

A number of ascidians including *Ascidia mentula* and *Phallusia mamillata* appear to thrive in fairly quiet, silty waters. The sponge *Suberites carnosus* is also found in such situations, together with the anemones *Sagartia troglodytes* and *Cereus pedunculatus*, which live with the column thrust through the layer of sediment, and attached to stones or rocks beneath. A slight increase in water movement enables a greater range of organisms to colonize sheltered areas, and species such as *Caryophyllia smithii*, *Anemonia viridis* and *Alcyonium glomeratum* may be found. *A. glomeratum* is an octocoral with branching, finger-like lobes, and is restricted to sheltered, low-energy situations. Its close relative *A. digitatum* has strong, stout lobes and can withstand water movements over 100 cm sec^{-1}. The species of kelp typical of sheltered localities is *Laminaria saccharina*, and other algal species include *Halidrys siliquosa* and *Dictyota dichotoma*.

Theoretically, because the biota of sheltered sites is not subjected to strong water movements, algae and animals are much more likely to be able to survive from year to year than they would be in exposed conditions where they are damaged by winter storms. However, in extremely sheltered locations there may be other reasons why communities are disrupted. For example, if the water becomes stratified during summer an oxycline may develop, resulting in severe oxygen depletion for the benthic fauna. Extreme conditions such as this may develop in sea lochs, and have been described for Lough Ine in Eire and Abereiddy quarry, a flooded seawater quarry in Wales (Hiscock and Hoare, 1975; Hiscock, 1983). In both places, stagnant conditions and the seasonal destruction of the benthic fauna appear to be a regular occurrence.

4.3.2 Semi- to moderately-exposed sites

There are many coastal and offshore sites where tidal currents are between about 20 and 100 cm sec^{-1} (0.5–2 knots), and where water movement from wave action is of a similar velocity for the greater part of the year. Such conditions suit a wide range of algae and animals and generally lead to the development of rich and extremely diverse communities. Some species have a fairly wide tolerance and are found in a variety of situations; others have more distinct preferences.

One of the features of shallower zones of moderately exposed rocky sites are the kelps. *Alaria esculenta* (especially on northern shores) and *Laminaria digitata* are often common around the low water mark, while the dominant species slightly deeper is *Laminaria hyperborea*. Under optimal conditions this species may grow 2–3 m high and form a dense forest showing distinct vertical layering. On the sea-bed itself are the holdfasts, sometimes packed so closely that they are touching. They and any rock spaces between them are colonized by organisms forming either crust, turf or meadow, depending on the environmental or biological factors operating at that particular site. The encrusters that do best in more exposed conditions are often calcareous algae such as *Lithothamnion*, sometimes bryozoans or ascidians. The turf layer, which may be 10 cm or more high, consists of hydroids, erect bryozoans or small algae. The appearance of the turf may alter with the season since many of the constituent species have annual or shorter life spans. Amongst the turf are large sessile organisms such as sponges and anemones, together with an array of mobile animals including isopods, amphipods, worms, nudibranchs, small crabs, gastropods, starfish, sea urchins and fish such as scorpionfish, blennies and gobies (Fig. 4.3).

Where the forest is more open and light penetration is relatively high the spaces between the holdfasts may be dominated by species forming a lush and distinct meadow. For example the red alga *Delesseria sanguinea* is common in these situations, and often itself becomes overgrown with a range of small sessile organisms as the summer progresses. The mid-layer is occupied by the stipes themselves, and their epiphytic growths (p. 73) and the upper layer by the canopy of overlapping blades. Wrasse often swim between the stipes, while Pollack and other members of the cod family hover just above the canopy.

Under ideal conditions kelp plants may live for 10 years or more, although the blade is shed annually and only the holdfast and stipe are perennial. This annual cycle of change is especially evident in early spring when the new blade

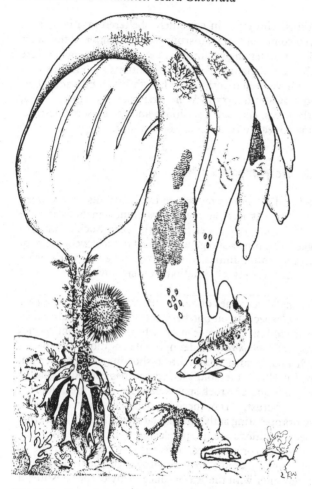

Fig. 4.3 Kelp, *Laminaria hyperborea*, as a habitat.

is making rapid growth, and the old encrusted blade is on the verge of becoming detached. The undergrowth also changes with the season as annual species such as many of the hydroids and foliaceous algae come and go.

In deeper areas, or in overhanging, vertical or other shaded places, animals rather than algae predominate. Sometimes dense stands of a single species occur covering rock surfaces several metres across. The hydroid *Tubularia indivisa*, the Jewel Anemone *Corynactis viridis*, Plumose Anemone *Metridium senile* and the soft coral *Alcyonium digitatum* are typical examples and probably owe their success to their ability to spread by means of asexual reproduction. A few mobile species aggregate in this way, for example the feather-star *Antedon bifida* on vertical rock surfaces and the brittle-star *Ophiothrix fragilis* on flat bedrock (p. 70). In other areas there is a mixture of species, often including the anemone *Actinothoe sphyrodeta*, the ascidian *Clavelina lepadiformis*, and various

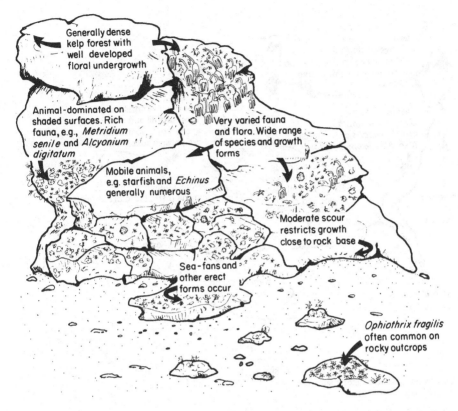

Generally dense kelp forest with well developed floral undergrowth

Animal-dominated on shaded surfaces. Rich fauna, e.g., *Metridium senile* and *Alcyonium digitatum*

Very varied fauna and flora. Wide range of species and growth forms

Mobile animals, e.g. starfish and *Echinus* generally numerous

Moderate scour restricts growth close to rock base

Sea-fans and other erect forms occur

Ophiothrix fragilis often common on rocky outcrops

Fig. 4.4 Characteristic features of a rocky subtidal site subjected to moderate water movement.

sponges. Many of the sessile organisms in these areas are suspension feeders, a few are deposit feeders, or can switch from one type of feeding to the other according to the availability of food. There are also grazers, scavengers and predatory species, including molluscs, echinoderms, crustaceans and fish.

There is considerable variation in the amount of silt and sand present on the sea-bed in moderately exposed areas, which relates to the precise degree of water movement and the availability of these sediments. The cup coral *Caryophyllia smithii* and erect sponges such as *Polymastia boletiformis* are examples of organisms that live in exposed conditions yet can withstand some siltation. The alga *Grateloupia dichotoma*, the Dahlia Anemone *Urticina felina* and the hydroid *Hydrallmania falcata* are amongst the species that can tolerate sandy or silty conditions, and the sponge *Ciocalypta penicillus* is found specifically on sand-coated rocks.

4.3.3 Very to extremely exposed sites

Sites where either current of wave-induced water movement is consistently

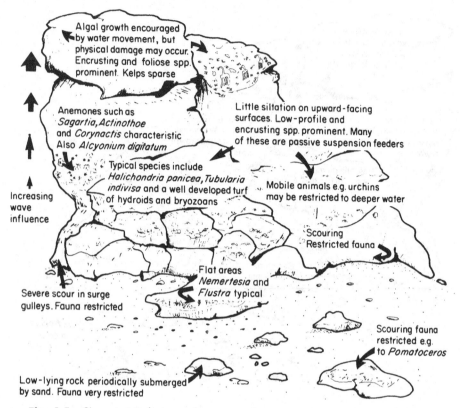

Algal growth encouraged by water movement, but physical damage may occur. Encrusting and foliose spp. prominent. Kelps sparse

Anemones such as *Sagartia,Actinothoe* and *Corynactis* characteristic Also *Alcyonium digitatum*

Little siltation on upward-facing surfaces. Low-profile and encrusting spp. prominent. Many of these are passive suspension feeders

Typical species include *Halichondria panicea, Tubularia indivisa* and a well developed turf of hydroids and bryozoans

Mobile animals e.g. urchins may be restricted to deeper water

Increasing wave influence

Scouring Restricted fauna

Flat areas *Nemertesia* and *Flustra* typical

Severe scour in surge gulleys. Fauna restricted

Scouring fauna restricted e.g. to *Pomatoceros*

Low-lying rock periodically submerged by sand. Fauna very restricted

Fig. 4.5 Characteristic features of a wave and current-exposed rocky subtidal site.

above 100 cm sec⁻¹, and may be much higher, expose benthic communities to considerable physical stress, and in these situations it is to be expected that there will be fewer, well-adapted species. The extent to which organisms can live in these areas depends on the precise type and strength of water flow, the degree of scouring and the amount of shelter provided by contours of the sea-bed and by the biota itself.

Alaria esculenta, found in northern areas, is the kelp best adapted to shallow areas exposed to very strong wave action, but kelp plants in general, if they manage to maintain a presence at all, are small. This is because they are plucked from the rocks before they have a chance to grow to full maturity. Soft-bodied animals are also considerably affected and the most successful are tough, low-growing or encrusting species such as the sponge *Halichondria panicea*, barnacles and encrusting bryozoans. Erect species can also survive under exposed conditions provided they are sufficiently flexible and reduce drag by having a shape that enables the water to flow past with little friction. The bryozoan *Flustra foliacea* and the hydroid *Nemertesia antennina* are both adapted in this way. It is difficult for mobile animals to maintain a foothold under conditions of extreme water movement, especially species such as gastropod molluscs, sea-

above 100 cm sec^{-1}

urchins and starfish that are neither hydrodynamically suited to fast water flow, nor particularly adept at gripping onto the sea-bed. Crabs and other crustaceans are more successful, partly because of their habit of wedging themselves into crevices.

Steeply sloping and vertical rock faces adjacent to sandy areas and subjected to strong water movement show a distinct zonation from base to top due to the effects of scouring. This is seen clearly in sandy surge gullies where rock surfaces at the very base of the gulley are devoid of sessile organisms, while a few centimetres higher up sessile organisms are present but are restricted to resistant species such as barnacles, encrusting bryozoans, crustose red algae (e.g. *Lithothamnion* and *Pseudolithoderma*) and the keel-worm *Pomatoceros triqueter*. A metre or so up from the bottom a richer community begins to develop, typically including cup corals *Caryophyllia smithii*, *Metridium senile* (Fig. 4.6) and *Alcyonium digitatum*. Two species characteristic of many surge gullies are the small red ascidian *Dendrodoa grossularia*, and the calcareous sponge *Clathrina coriacea*.

Deeper, flatter areas of rocky sea-bed subjected to strong tidal flow and scouring also have an impoverished fauna. In parts of the Bristol Channel boulders and bedrock surfaces are worn smooth and the only refuge for mobile species is in the burrows of boring bivalves such as *Pholas dactylus* and *Hiatella arctica* (Warwick, 1984). Tide-swept rocky substrata at a depth of 50–55 m in the central English Channel that are affected by severe scour and periodic deep submergence by sand and gravel are found to be colonized mainly by barnacles and *Pomatoceros triqueter*. The small hermit crab *Pagurus pubescens* is also characteristic (Holme and Wilson, 1985). In adjacent areas where scouring and submergence is not quite so pronounced, the bryozoan *Flustra foliacea* is

Fig. 4.6 The Plumose Anemome, *Metridium senile*, typical of current-exposed situations.

characteristic; sponges are absent, and the only anthozoan is the Dahlia Anemone *Urticina felina*, which is well adapted to such conditions, and is often seen with the column projecting through the sand and the base attached below to rock or cobble. Where scouring and submergence are even less of a problem the community is more diverse. The ascidian *Polycarpa violacea* is characteristic and a variety of hydroids, anemones, bryozoans and sponges are present, although only *Dysidea fragilis* is common. Finally, on the surface of immobile hard bottoms of pebbles, cobbles, boulders and rock outcrops not subject to scour or submergence there is a rich and stable assemblage of organisms characterized by a diverse sponge cover. Included are the sponges *Hemimycale columella* and *Myxilla incrustans*, hydroids such as *Nemertesia ramosa*, *N. antennina* and *Sertularia* spp., the anthozoans *Actinothoe sphyrodeta*, *Corynactis viridis*, *Sagartia elegans* and *S. troglodytes*, the bryozoans *Pentapora foliacea* and *Omalosecosa ramulosa*, the ascidians *Ascidia mentula*, *Botryllus schlosseri* and *Distomus variolosus*, the polychaetes *Polydora ciliata* and *Salmacina dysteri*, molluscs such as *Calliostoma ziziphinum* and *Ocenebra erinacea* and the echinoderm *Henricia sanguinolenta*.

There are many other situations where the flora and fauna are restricted by, and related to, the degree of water movement. For example in Lough Ine the sessile fauna on boulders in the rapids is clearly different from that in adjacent areas sheltered from current. Limits of tolerance are found to vary from species to species. The limpets, *Patella* species are most tolerant, followed by *Mytilus edulis*, which is also resistant to current but does not colonize sheltered bays. In contrast, *Corynactis viridis* appears to find the current on the sill, which reaches 300 cm sec⁻¹, too strong, and this species is restricted to the underside of boulders, and also penetrates some way into sheltered areas. The Plumose Anemone *Metridium senile* prefers the main stream, as do various hydroids (e.g. *Sertularia operculata* and *Plumularia setacea*) (Lilly *et al.*, 1953).

4.4 Artificial substrata

Sometimes the communities associated with wrecks and artificial structures are very similar to those on the sea-bed around them, but much depends on the type and stability of the structure involved, the length of time it has been in the water, and its precise location. Most artificial substrata, and especially those intentionally constructed in the sea, such as jetties, harbour walls and oil rigs are stable, but shipwrecks vary greatly in their stability, depending on where they have landed. Often they are in exposed locations, and are being continually moved around and broken up. However, eventually, a relatively stable condition is reached.

Wrecks which have dropped onto soft substrata, or oil rigs purposely placed there, act as miniature artificial reefs, and enable hard-bottom species to spread into areas where they would not normally be present. These communities in particular are often noticeably different from communities of the nearest natural rocky substrata, perhaps because the plankton passing by does not include such a wide range of potential colonists. In this respect, the length of time that larvae spend in the plankton, and the speed and direction of the water currents in which they are carried is clearly a major influence. Species such as *Tubularia larynx*, with its non-swimming larva which soon sinks to the sea-bed, and *Ascidiella aspersa*, which settles after only a few hours (Goodman and Ralph, 1981), are unlikely to reach isolated structures. It has been found, for example, that it may take over two years before a sizeable fouling community develops on artificial structures placed beyond the 50 m depth contour in the North Sea (Fig. 4.7). In addition, it will be difficult for adult benthic grazers and predators to reach these islands of hard substrata, and this will alter the types of biological pressure to which the sessile biota is subjected, in the initial phases at least.

It is also not unusual to find differences even when the artificial structure is lying on or immediately beside a natural hard substratum. This may be due to a combination of factors, including peculiarities in water flow around the structure, and increased habitat complexity. A particularly good illustration of this is that the Jewel Anemone *Corynactis viridis* is present on wrecks off the Sussex coast, yet absent from surrounding rocky areas. Their presence on these artificial structures marks the easterly boundary of their distribution along the English Channel, and it seems that the wrecks provide some special feature(s) absent from the natural hard substrata.

A feature of many wrecks and artificial structures is the fish life they support. Tompot Blennies, wrasse and Conger Eels find refuge amongst broken-up structures, and the Bib *Trisopterus luscus* is especially characteristic of wrecks situated in moderately exposed areas. They often occur in dense shoals in these situations, together with Pollack, *Pollachius pollachius*.

4.5 Unstable hard surfaces

In areas of extremely strong water movement small rocks, cobbles and pebbles are almost constantly on the move, and as they roll against each other, they are scoured clean. However, pebbles in locations subjected to less extreme conditions may be stable for months at a time, especially during summer, and become colonized by transient opportunistic species such as barnacles, encrusting bryozoans and foliose algae. The calcareous tube-worm *Pomatoceros triqueter* also does well in these situations, often colonizing the undersides of pebbles while algae dominate upward-facing surfaces (Fig. 4.8).

Some longer-lived animals including *Urticina felina* and hermit crabs *Pagurus* spp. also withstand these conditions, and polychaetes such as *Polycirrus* can colonize the spaces between the pebbles. As the population of attached organisms builds up so the drag increases, making it easier for the pebbles to be rolled around or lifted from the sea-bed. If this happens the colonists are often dislodged, but they may remain in place, having acted like a parachute in lifting and floating the pebble from place to place. Algae such as *Chorda filum* and *Laminaria saccharina* do this, while on the south coast of England the recently introduced Japweed, *Sargassum muticum* with its buoyant vesicles is even more effective. Not surprisingly, larger pebbles in areas of stiller water support attached assemblages with higher drag characteristics either than smaller ones, or those subjected to strong water movements.

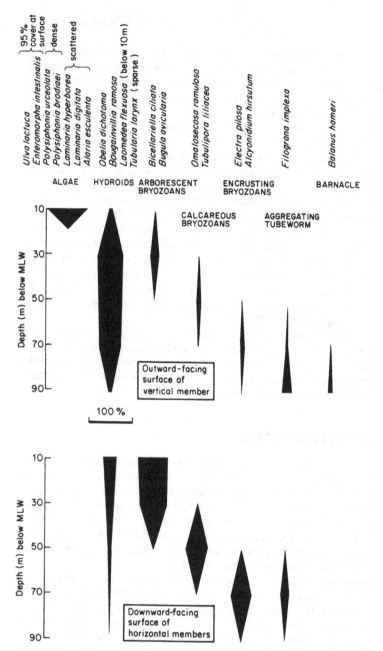

Fig. 4.7 Zonation of marine organisms on the Montrose jacket in 1980, five years after it had been installed 195 km east of Peterhead in north-east Scotland, in 90 m of water. Percentage cover (estimated from analysis of colour photographs), *(top)* on outward-facing surface of the vertical legs. *(bottom)* on downward-facing horizontal members. From data in Forteath *et al.* (1982). Ubiquitous species not sufficiently abundant to be included in the per cent cover analysis included the tubeworms *Pomatoceros triqueter* and *Hydroides norvegica* (mostly overgrown by bryozoans, and algae above 31 m); the sponge *Leucosolenia complicata* (rare below 51 m); the amphipod *Jassa falcata* and nudibranch *Dendronotus frondosus* (amongst hydroid samples), and the errant polychaete *Nereis pelagica* (widespread).

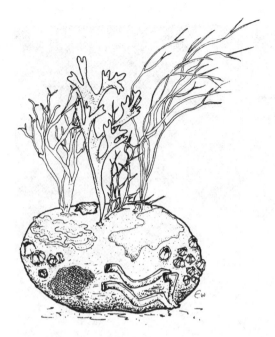

Fig. 4.8 Pebble with annual community. Well-lit upper surface with red filamentous algae (from left *Grateloupia filicina, Chondrus crispus, Gigartina acicularis* and *Solieria chordalis*), and crustose spp. (*Peysonnelia, Lithophyllum* and *Hildenbrandia*). Lower surface with barnacles, the bryozoan *Electra pilosa* and the tube-worm *Pomatoceros triqueter*.

4.6 Biotic substrata

A significant proportion of sessile benthic epifauna is attached to other plants and animals, rather than directly on to the rock. However, this is not a universal phenomenon, for there are many organisms which consistently maintain the surface of their bodies, or their shells, free of 'fouling' organisms. Sponges, echinoderms, bryozoans and most tunicates have mechanisms to rid themselves of settlers, which would cause problems if they did settle, because respiratory and feeding devices would become obscured or be interfered with. However, it seems that where a non-dynamic surface is available that would cause no problems to the animal or plant, then this may be colonized.

The test of large tunicates is often colonized by anemones (Fig. 4.9), other tunicates and sponges. Large hydroids such as *Nemertesia* support a variety of sessile organisms, as do kelps and other large algae. The shells of many molluscs and crustaceans provide a hard substratum for barnacles, sponges, hydroids, bryozoans and filamentous algae in particular.

It is also common to find mobile species using other organisms as a perch. For example the brittle-star *Ophiothrix fragilis* is often found nestling between the lobes of *Alcyonium digitatum*, with its arms extended to catch passing

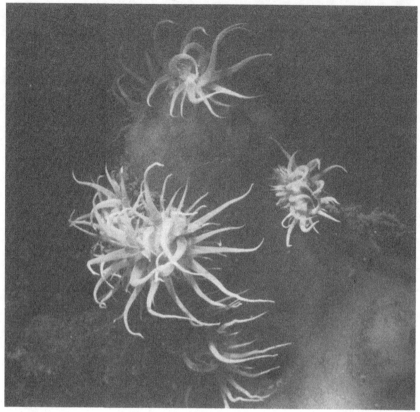

Fig. 4.9 The tunicate *Phallusia mamillata*, with epizoic anemones, *Actinothoe sphyrodeta*.

suspended material. In a similar way caprellids cling to the stems of the hydroid *Nemertesia*, so stationing themselves in an advantageous position for capturing passing prey.

Some of this settlement is both complex and predictable, and distinct communities may develop as described in Chapter 7.

4.7 Crevices and holes

Crevices, because they are such distinct microhabitats, warrant special mention, especially as they are found in virtually every hard-bottom habitat. They occur between natural rock faults, under and between boulders or the struts on a wreck, as well as between the irregular protrusions of sessile invertebrates. Conditions in crevices are often distinctly different from those in the general surroundings because they are shaded, sheltered from water move-

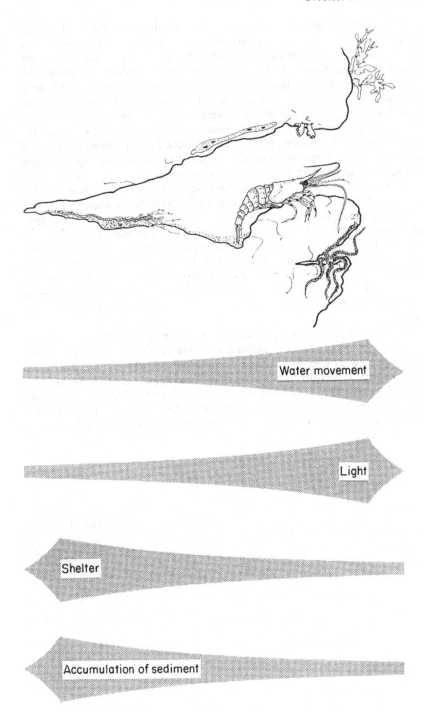

Fig. 4.10 Crevice microhabitat.

ment, scour and large predators, and serve as silt traps (Fig. 4.10). Being shaded from the light, unless upward-facing and in shallow water, crevices are the domain of animals, even though adjacent open surfaces may be entirely algal-dominated. Some of the inhabitants are invariably associated with crevices, either being partially or completely hidden, or moving regularly in and out to feed. Others are less dependent and also occur in various other situations. The groups most commonly represented are worms, small crustaceans, brittle-stars, sea-cucumbers and fish. Some of the worms feed on silt which has accumulated in the crevices, but the other residents either extend some sort of feeding apparatus beyond the crevice entrance to trap suspended organisms and detritus, or dart out bodily to capture passing prey.

In the former category, for example, are brittle-stars such as *Ophiopholis aculeata* and *Ophiactis balli* both of which are found almost exclusively in rock crevices or amongst sessile invertebrates. The disk is concealed but the arms protrude and rows of small tube feet are used to ensnare food. The sea-cucumbers *Pawsonia saxicola* and *Aslia lefevrei* (widely referred to as *Cucumaria normani*) are common in crevices in the south-west of the British Isles, and have conspicuous branched tentacles that protrude from the end of the soft, gherkin-shaped body. Items of food are trapped on special adhesive areas on the extremities of the terminal tentacle branches, and then transferred to the mouth.

Typical amongst the larger crustaceans found in crevices are squat lobsters *Galathea* spp. and a variety of shrimps and prawns. These capture planktonic food that is swept into the crevices, or move out periodically into more open areas to feed. Fishes invariably associated with crevices include the Leopard Spotted Goby, *Thorogobius ephippiatus*, and Yarrell's Blenny, *Chirolophis ascanii*. Little is known about the behaviour of these fish, for example whether they are territorial, and how far and frequently they range from their crevices.

5

Benthic Communities: Soft Substrata

5.1 Study methods

Traditionally, soft sea-bed communities have been investigated by remote sampling using grabs, dredges and corers, and this type of gear is still widely used, especially in deeper areas. This is then followed by extraction of the animals by sifting the sediment through meshes of various sizes. In recent years an increasing amount of quantitative work in shallow areas has been carried out directly by scuba divers. Photographic surveys of the soft sea-bed using television and other cameras are also providing useful information on the distribution of surface-dwelling animals. Relatively few ecological and behavioural studies have been carried out on soft sea-bed communities although some interesting techniques have been employed; for example the use of quick-setting resins to map the structure of large burrows. Even with advances such as this it is difficult to carry out *in situ* studies of sediment communities, simply because much of the fauna is hidden from view. However, some infaunal species appear at times on the surface or leave evidence of their presence which provides information about their distribution and behaviour.

5.2 The habitat and its inhabitants

Sediments consist of particles ranging from gravel (maximum diameter 4 mm) to clay (less than 0.002 mm diameter). Although the particles may initially be angular or irregular, they gradually become rounded due to abrasion. Particulate material is inherently unstable, and provides few opportunities for sessile organisms to become attached to its surface. Algal growth is generally restricted to micro-algae, and communities are dominated by animals living within the sediment (infauna), or moving about on its surface (mobile epifauna). Animals that live within the sediment may simply bury themselves (heart-urchins), or they may push their way through it without forming a burrow (bivalve molluscs). Others form distinctive burrows, especially in mud. In some respects the division between infauna and epifauna is not a good one because of the animals that either burrow during the day and emerge at night to feed, or live partially buried, generally with the main part of the body in the sediment, and the feeding structures on or above the sediment surface.

Often the fauna is split for convenience into categories according to size. This classification came into use to describe the fauna collected by different sampling or sorting techniques, but the categories also have certain biological features. Large, conspicuous animals, down to those retained by a 0.5 mm mesh, comprise the macrobenthos, although some workers use a 1 mm mesh to separate the macrofauna. Some of the macrofaunal animals, including anemones, starfish, hermit crabs, shrimps, prawns and fish spend much of their time on the surface of the sediment. Others, principally polychaetes, bivalve molluscs, echinoderms and small crustaceans, burrow into it or live within the interstitial spaces in coarse deposits.

Animals small enough to pass through a sieve with a mesh size of 0.5 mm but large enough to be retained by a mesh of 0.062 mm are referred to as meio-benthos. However, this is not necessarily the minimum dimension of the animal, since many are elongate and thus will slip through the mesh. Members of the meiobenthos are neither large nor strong enough to burrow through the sediment, but live within it, often in the interstitial spaces between particles. Nematodes are often the dominant group, with several hundred occurring per square centimetre. Many of these small worms are deposit feeders, others graze on organisms attached to the sediment particles, or prey on other interstitial animals. Copepods, especially harpactacoids, are also abundant, and specific communities have been recognized for different grades of sediment. Archian-nelids, small polychaete worms specially adapted for living in interstitial spaces, are often abundant and many other groups are represented, including turbellarians, gastrotrichs, tardigrades, rotifers and some of the larger proto-zoans, such as foraminiferans and ciliates. There is also a significant temporary population of larval and juvenile members of the macrofauna which pass through the meiofaunal category. Despite the small size of its individual members, the meiobenthos is immensely important in the ecology of the soft sea-bed.

Finally there is the microbenthos, including bacteria, protozoans, and micro-algae which pass through a 0.062 mm mesh. The most important micro-algae in soft-bottom habitats are the episammic diatoms. These have very limited motility, and accumulate in interstitial spaces and on the sediment surface. Few studies of microbenthic communities have been made, although the role of bacteria in the recycling of nutrients is well accepted.

5.3 Abiotic and biotic factors influencing community structure

5.3.1 Sediment type and degree of sorting

A universal method of classifying sediments is by grain size, and there is a corre-lation between this physical scale and the associated biological communities. However, this is not as simple as it might appear because sediments are seldom uniform in composition and may vary over distances as small as a few cen-timetres. This happens, for example, where the sediment is thrown into ripples by water movement, or into mounds by the activities of burrowing animals, and sediment grade alters from the peak of a ripple or mound to its trough or base.

Species diversity, as well as overall community structure, is influenced by the

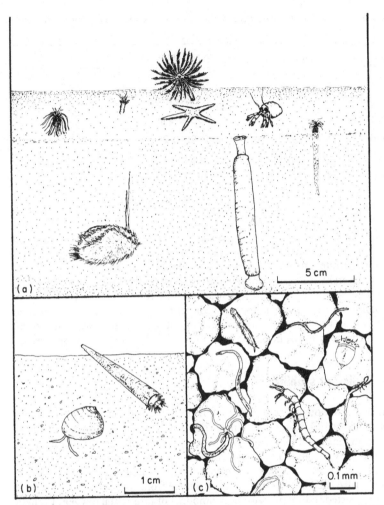

Fig. 5.1 Diagrammatic representation of the fauna of sediments. (**a**) Epifauna (starfish, *Astropecten irregularis* and hermit crab, *Pagurus* sp.), partially buried macrofauna (*Cerianthus, Sabella* and other tubeworms) and burrowing macrofauna (*Echinocardium* and *Ensis*). (**b**) Small infaunal deposit feeders (the polychaete *Lagis koreni* and the bivalve *Nucula* sp.). (**c**) Meiofauna (left to right: hydrozoan, nematodes, gastrotrich, harpacticoid copepod, rotifer, polychaete).

sediment type. Many species are found in or on a range of sediment types, but others have a more restricted distribution. The bivalve *Nucula* illustrates this point well, with each of the five species showing a distinct preference for a particular grade of sediment. The balance between the infaunal and epifaunal element also varies according to sediment grade. Coarse sediments, which are unstable and difficult to burrow into, are dominated by epifauna, while finer sediments are increasingly dominated by infauna. The greatest diversity of

Table 5.1   Grain size classified according to the Wentworth scale (a geometric scale suitable for statistical treatment).

	Grain diameter
Gravel	2–4 mm
Coarse sand	0.5–2.0 mm
Medium sand	0.25–0.5 mm
Fine sand	0.125–0.25 mm
Very fine sand	0.063–0.125 mm
Silt	0.002–0.063 mm
Clay	0.00006–0.002 mm

macro-infaunal species is generally associated with poorly-sorted sands because they are physically heterogeneous, reasonably stable, relatively easy for burrowers to penetrate, and contain a supply of deposited organic matter. The density and diversity of meiobenthos follows a slightly different pattern but generally both decrease as the interstitial spaces become smaller. This happens in fine sands, and also in poorly-sorted sediments where small grains make their way into the spaces between larger one. In mud meiobenthic organisms occur but lie in the mud itself, because there are virtually no interstitial spaces.

5.3.2 Sediment stability

Sediments in shallow water are affected by both wave action and tidal currents, but below about 100 m any sediment sorting which occurs is from the effects of water movement from tidal flow (see page 7). Sediment can also be disturbed from within, by the animals themselves, but these two types of disturbance do not necessarily work in concert, or have the same outcome in terms of community structure. Persistent physical disturbance from water currents results in clean, well-sorted sediments with high amounts of interstitial oxygen. Periodic disturbance from storms leads to upturning and washing-out of both silt and fauna. Sediments in areas of sea-bed unaffected by water movement are less well sorted, and may have substantial admixtures of silt and organic matter which favours deposit feeders of all types.

The fauna may itself alter both the stability and the nature of the substratum, and so affect community structure. Unselective deposit feeders, including the polychaete *Arenicola marina* and the spoon worm *Maxmulleria lankesteri*, ingest particulate material and re-sort the sediment. Predatory animals such as catworms *Nephtys* spp. and the gastropod mollusc *Natica* also loosen it and so make it more readily penetrated by burrowing animals. These types of disturbance decrease the stability of the sediment and mean that it is more easily overturned during storms, and a more difficult environment for juveniles to settle. Conversely, tubes of worms such as *Lanice conchilega* and *Melinna cristata*, which are tough and often densely packed, help to consolidate the sand, so hindering the activities of predatory burrowers, but enabling other sedentary animals to establish themselves.

5.3.3 Light, depth and profile

Light plays much less of an obvious role than in hard-bottom habitats because macro-algae can rarely anchor themselves on particulate substrata, and so the algal-based communities and related depth zonation patterns so typical of hard substrata do not develop. This does not mean that community structure remains constant with depth, merely that factors other than light are involved. Similarly, spatial differences in communities related to changes in profile, which are another feature of hard-bottom communities, are much less prevalent on particulate substrata because of their relative uniformity. In contrast, a feature of sediment communities is that there is often a distinct zonation pattern from the surface downwards which follows the gradation in physico-chemical characteristics.

5.3.4 Organic content

The organic content of sediments comes from two major sources. The bulk is derived from external sources, but some is generated from within by the activities of the animals themselves. The amount that settles depends on the amount washed into an area or dropping down from above, and the degree of water movement. Coastal waters, especially in the vicinity of estuaries, tend to be loaded with suspended organic matter, and this will settle when water movement is reasonably slow. Fine particles are deposited at current speeds of around 5 cm sec^{-1}, but may be resuspended by speeds of around 15 cm sec^{-1}. However, once they are well settled resuspension is less likely because of cohesion of the particles. Organically rich sediments often support rich faunas dominated by deposit feeders, but they also encourage considerable bacterial activity, and this can lead to the development of anoxic conditions fairly close to the sediment surface and to a reduction in faunal diversity.

The picture below of the faunal characteristics of different areas of soft sea-bed is inevitably only a broad outline, partly because of the almost infinite variations that occur in the substratum. A small shift in the proportion of sand to silt, or the admixture of coarse sediment can lead to corresponding changes in the abundance and diversity of the fauna and in species composition. Conversely, there are some species, including various skates, rays and flatfish, that are associated with a wide variety of sediment types.

5.4 Coarse sediments

Coarse deposits in areas where water movement is relatively vigorous are well sorted and have a low organic content because fine material cannot settle. For example much of the bed of the English Channel is covered by coarse sand and gravel and where these deposits are highly mobile there may be an almost complete lack of infauna, and few epifaunal species. However, where there is more stability then diversity increases. Suspension feeders rather than infaunal deposit feeders are dominant, and echinoderms are often well represented. Brittle-stars such as *Ophiothrix fragilis* and *Ophiocomina nigra* may occur in dense beds on the surface, while *Ophiopsila annulosa* and *Amphiura securigera*, both south-western species, have the disk dug into the gravel, and the arms protruding to catch suspended material.

Echinoids typically found in gravel are the Yellow Sea-potato, *Echinocardium flavescens*, the Purple Heart-urchin, *Spatangus purpureus*, and the tiny (1 cm long) *Echinocymus pusillus*. The only holothurian to be found in coarse, mobile gravel is *Neopentadactyla mixta*, which has conspicuous white tentacles that emerge periodically at the surface. Bivalves such as *Nucula hanleyi*, *Glycymeris glycymeris*, *Venus fasciata*, and *Ensis arcuatus* are typical of coarse gravel.

In areas where relatively coarse deposits occur, but water movement is only slight, if silt is present in suspension it will be deposited and accumulate on the surface of the gravel and in interstitial spaces. This leads to the development of a more diverse fauna than that associated with clean gravel. Most noticeable is the increase in numbers of deposit feeders. In the English Channel there are distinct muddy gravel communities. The most common infaunal species are burrowing crustaceans such as *Upogebia deltaura* and *U. stellata*, the bivalves *Nucula nucleus* and *Venus verrucosa*, and the gastropods *Turritella communis* and *Gibbula magus*.

In addition, a sessile epibenthic community may develop. Small bryozoans and hydroids manage to colonize gravel if conditions are calm, and larger hydroids such as *Nemertesia* can develop extensive fibrous rooting systems which enables them to maintain a hold and also binds the sediment together. The number and variety of mobile epifaunal species also tends to increase with increased physical diversity of the sediment surface. Mixed coarse sediments with some silt present generally support animals such as scale-worms (e.g. *Lepidonotus squamatus*), crustaceans (e.g. *Galathea*, *Macropodia*, *Inachus* and *Liocarcinus depurator*). The whelk *Buccinum undatum* and other molluscs such as *Nassarius reticulatus*, *Gibbula magus*, *Turritella communis*, *Pecten maximus* and *Chlamys* spp. are also common on this type of sea-bed. Starfish such as the sunstar *Solaster papposus*, the Goose Foot Starfish *Anseropoda placenta*, and *Luidia ciliaris*, an unselective feeder on other echinoderms, are typical. The brittle-stars *Ophiocomina nigra* and *Ophiura albida* and the small urchin *Psammechinus miliaris* may also be common. A variety of fish occur on mixed, gravelly sediments, including the dragonet *Callionymus lyra* and several species of flatfish.

5.5 Clean, well-sorted sands

Clean sands occur in exposed inshore areas or in high-energy offshore situations away from coastal silt input or where currents are strong enough to prevent accumulation of fine sediment. The biota is shaped by physical rather than biological forces. An extreme example of a physically stressed sandy-bottom area of the sea-bed is in the southern North Sea where there is a series of sand-banks about 25 m high, 15–25 km long and 3–6 km wide, separated by channels 4–6 km wide. The sand on each of the banks is thrown into sand-waves about 5–6 m high, on which are megaripples about 60 cm high and ripples about 2 cm high, but these surface sands are constantly shifting as a result of the very strong tidal currents. At certain times, particularly during storms, the whole top of the sand-bank is destroyed, then rebuilt when conditions are calmer. A characteristic of the sand-bank system is that sessile tube-building polychaetes are represented by a small number of individuals only, whereas the dominant forms are mobile and quickly burrowing animals such as crustaceans and the polychaetes *Nephtys cirrosa*, *Hesionura augeneri* and *Microphthalmus listensis*. These animals are able to withstand physical disturbance and to re-enter the sediment rapidly after being washed out. The sediment in the sand-bank systems ranges from fine to coarse clean sands, and the density of individuals is highest in the coarsest grade, mainly due to large numbers of interstitial polychaetes. The bottom of some of the sand-waves is enriched with mud, and characterized by an abundant fauna (Vanosmael *et al.*, 1982). Another dune system has been described on a coarse sand bottom off the Isle of Man, where there are waves up to 12 m high and 330 m long which shift by as much as 74 cm during a single tide. This instability has led to an impoverished infauna, but dense shoals of sand eels were present. An extensive system of sand-banks and waves is also present on the continental shelf at the south-western edge of the Celtic Sea, while sand-waves up to 20 metres in height occur in the approaches to the Irish Sea and the English Channel.

The benthic community associated with well-sorted inshore sands off the west coast of Scotland is similar to that described for offshore sites although the environment is less stressed (McIntyre and Eleftheriou, 1968). Polychaetes were dominant with *Nephtys cirrosa*, an infaunal species characteristic of clean sands, most widely distributed. The majority of macro-crustaceans were amphipods, with *Bathyporeia* dominant, but isopods and cumaceans were also present. Amongst molluscs, *Tellina* spp. were conspicuous. There was a range of epibenthic animals including shrimps (*Crangon crangon*), hermit crabs (*Pagurus bernhardus*), starfish (*Asterias rubens*) and fish such as pipefish (*Syngnathus acus*), weeverfish (*Trachinus vipera*), the Gunnel (*Pholis gunnellus*), and juvenile gadoids, plus plaice and dabs. Flatfish capture mobile epifauna and also unearth soft-bodied and shelled animals from the sediment, or clip off protruding siphons and tentacles with their teeth. In turn they are preyed upon by larger species such as the Monkfish *Squatina squatina*.

Species sometimes used as indicators of clean sand include the polychaete worms *Spiophanes bombyx*, *Eulalia tripunctata*, and *Glycera capitata*, the bivalves *Venus striatula*, *Cerastoderma echinatum*, *Spisula elliptica* and *Ensis siliqua*, the gastropod *Natica catena*, the brittle-stars *Ophiura albida* (on the surface) and *Amphiura brachiata* (burrowing). The heart-urchin *Echinocardium cordatum* and the Masked Crab *Corystes cassivelaunus* are also typical.

5.6 Poorly-sorted, silty sands

Poor-sorting and siltiness tend to go together, because both occur in areas

Fig. 5.2   The gastropod mollusc, *Natica alderi*, a predator typically found in sandy sediment.

where water movement is insufficient to sift the sediment and sort the fine particles from the coarse, or to prevent silt and detritus from settling. Typically, muddy sand communities are dominated by deposit feeders, which rework the sediment and transform mud particles into faecal pellets. There are also many carnivorous animals, and these add to the disturbance of the sediment as they burrow through it in search of prey. A positive aspect of these disturbances is that they help to prevent anaerobic conditions from developing. Generally the biota is dominated by infaunal species, but in some areas dense beds of tube-dwelling worms develop as a turf on the surface. *Polydora* spp. often do this, and at a depth of 70–75 m in the Irish Sea, *Ampharete falcata* has been found at densities of around 3000 per square metre. Amongst this turf was a dense population of the cockle *Parvicardium ovale* (around 27 000 per square metre), and other bivalves such as *Nucula tenuis* and *Abra nitida* (Holme and Rees, 1986).

Polychaetes are often the numerically dominant macrofaunal group, but crustaceans and molluscs are also well represented, and the final balance varies from site to site (Fig. 5.3). In Galway Bay, 51 of the 85 macrofaunal species recorded from poorly-sorted sands were polychaetes, but bivalves were also common (Shin, 1981a). In fine silty sand off the Northumberland coast, at a depth of 80 m, eighteen species accounted for 90% of all animals, and 12 of these species were polychaetes. However, in terms of biomass the burrowing crustacean *Calocaris macandreae* was dominant, comprising more than 30% of total biomass (Buchanan and Warwick, 1974). At a depth of 45 m off the North Yorkshire coast the dominant faunal groups were polychaetes (54 species), bivalves (16 spp.), echinoderms (6 spp.) and amphipods (3 spp.) (Atkins, 1983).

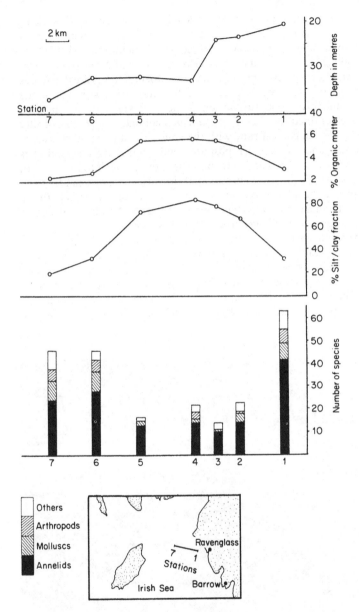

Fig. 5.3 Sediment features at 7 stations in the north-east Irish Sea, and numbers of infaunal species retained on a 0.5 mm sieve from 5 × 0.1 m Day grabs. Compiled from data in Jensen and Sheader (1986).

In muddy sands at Lyme Bay the fauna was dominated numerically by polychaetes, followed by bivalves and crustaceans. Over 80% of the macrofauna here were detrivores feeding at the sediment/water interface (Eagle and Hardiman, 1976).

Some muddy sand species are diagnostic even though they are not dominant, but there are variations in species composition from site to site, which match the range of environmental variables and biological interactions that occur. Amongst the polychaetes some, including *Glycera rouxi* and *Nephtys hombergi*, are predatory and burrow actively through the sediment. Others are detrivores or filter feeders, using palps and tentacles to collect food from within or on the surface of the sediment, or from suspension. Some species are relatively mobile, while others are confined to tubes. *Lagis koreni*, a common inhabitant of silty sands, builds a delicate conical tube with the anterior end opening below the sediment surface, and the smaller posterior end protruding above. It has prominent chaetae at the head end, which are used to burrow into the sand. In contrast, *Lanice conchilega*, which occurs both in silty and clean sands, has a fringed tube projecting a few millimetres above the surface, and uses both the fringe and the tentacles to collect suspended material. Another polychaete typical of muddy sand is the Lugworm *Arenicola marina* (Fig. 5.4).

Many of the molluscs in silty sand are infaunal bivalves, and these include species typical of shallow water such as *Ensis ensis, Macoma balthica* and *Mya arenaria*, and those such as *Abra alba* and *Nucula turgida* which are characteristic of deeper areas. Some of these infaunal bivalves are suspension feeders connected to the surface by their siphons, others have an elongate inhalent siphon that sweeps the sediment surface for detritus. Other molluscs include

Fig. 5.4   The Daisy Anemone, *Cereus pedunculatus*, protruding from muddy sand, and casts of the lugworm, *Arenicola marina*.

the opisthobranch *Philine quadripartita*, which ploughs through the sediment in search of prey.

The heart-urchin *Echinocardium cordatum* occurs in both muddy and clean sands, although it grows much more slowly in the former than the latter (Buchanan, 1966). It feeds by extending a group of tube feet to the surface to pick up sediment particles which are then transferred down to the mouth. As the sediment becomes siltier *E. cordatum* is replaced by another heart-urchin, *Brissopsis lyrifera*, which often occurs in sediments with a fairly high clay content. Various brittle-stars are associated with this type of sediment, and are considered as epifauna, although the disk is often buried. *Amphiura filiformis* is a characteristic species, which extends its arms up into the water column to feed on suspended material, while in *A. chiajei* the arms generally lie on the surface from which they collect deposited material. Holothurians such as *Paracucumaria hyndmani* and *Labidoplax digitata* feed in a similar way. One of the few echinoderms to occur on the surface of silty sand is the starfish *Luida sarsi*, which is found especially in deeper water.

Epifaunal organisms associated with silty sands are predominantly mobile species, including the crabs *Liocarcinus depurator*, *Atelecyclus rotundatus* and *Macropodia* spp. Predatory fish such as Dover Sole, *Solea solea* and members of the cod family also frequent these areas, but many also occur on coarser and mixed grades of sediment.

5.7 Muds

Muds consist of very fine particles of silt and clay and may be firm and consolidated, or loose and flocculent. They are typical of sheltered coastal localities such as lochs and estuaries on the one hand, and deeper, offshore areas on the other. Firm estuarine and coastal muds may support piddocks and the burrowing worm *Polydora ciliata*, while less well-consolidated muds are characterized by nereid, spionid and capitellid worms.

Extremely fine silty sediments present the fauna with various difficulties, partly because of the tendency for anaerobic conditions to develop and also because the silt clogs respiratory and feeding structures. This means that the fauna is restricted to the upper few centimetres and that there are few suspension feeders. Often the fauna shows low species diversity, even though biomass may be high, but much depends on the amount of silt present. In some parts of Sullom Voe for example the silt/clay content often exceeds 50%, and this type of sediment has low populations of epifauna, and the infauna consists almost entirely of small deposit-feeding polychaetes such as *Capitella* and *Lagis* (Addy, 1981). These animals rarely penetrate below 4 cm in these highly organic sediments where the oxygenated layer is only a few millimetres below the sediment/water interface. In other parts of the Voe the organic content is lower, and surface deposit-feeding terebellid and spionid polychaetes and the filter-feeding bivalve *Cochlodesma* are common. Here the upper oxygenated layer of sediment extends to between about 3 and 7 cm, but larger tube-dwelling deposit-feeding worms such as *Rhodine*, and the bivalve *Thyasira flexuosa*, which create extensive feeding channels in the sediment, frequently extend their activities to the deoxygenated zone (Pearson and Eleftheriou, 1981). Another bivalve typical of

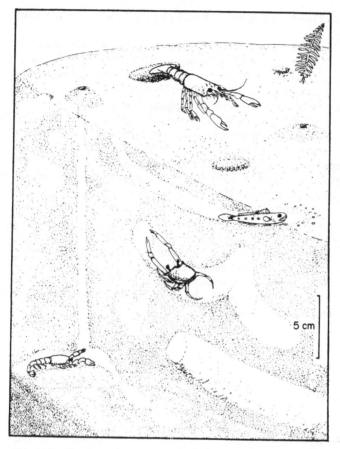

5 cm

Fig. 5.5 Diagram showing some inhabitants of subtidal mud and their burrows, including the crustaceans *Nephrops norvegicus*, *Goneplax rhomboides* and *Callianassa subterranea*, the goby, *Lesuerigobius friesii* and the anthozoan *Pennatula phosphorea*.

soft mud is *Nucula sulcata*. Echinoderms are less common in mud than in other sediments, except for the holothurians *Thyonidium commune* and *Thyone fusus*.

Muddy sediments are particularly suited to the construction of burrows, and these sometimes form a complex labyrinth beneath the surface. The main burrows of the Norway Lobster *Nephrops norvegicus* descend into the mud at a shallow angle and may consist of a single tunnel with an opening at each end, or may be more complex, with side branches and vertical shafts. The main tunnels, which lie roughly 20–30 cm below the sediment surface may be up to 10 cm in diameter, and often are a metre in length (Atkinson and Nash, 1985). The burrowing crab *Goneplax rhomboides* excavates with its chelae, and constructs smaller, narrower burrows. Even smaller burrows, about 1.5–3.0 cm in diameter and 20 cm in length, are made by Frie's Goby *Lesuerigobius friesii*, which carries mud to the tunnel entrance in its mouth. Other fish that burrow into mud include the Snake Blenny *Lumpenus lumpretaeformis* and the Red-band fish *Cepola rubescens. Cepola* burrows may be a metre deep, and the entrance is marked by a mound of sediment. Conspicuous volcano-shaped mounds with an entrance at the apex are indicative of burrowing, lobster-like crustaceans such as *Callianassa subterranea* and *Upogebia* spp.

The meiobenthic population of organically rich muddy sediments is almost invariably dominated by nematodes, followed by harpactacoid copepods. Nematodes thrive in oxygen-starved zones several centimetres from the surface, whereas other taxa tend to be concentrated in the very top layers, seldom penetrating deeper than about 2 cm. This situation does not always occur. For example, in certain areas in the northern North Sea, at depths of around 100 m, on very fine sand and mud there is an abundance of tubiculous suspension feeders, together with dense populations of foraminiferans such as *Astrorhiza limicola*. The pseudopodia of these protozoans can ramify over the sediment surface and through the interstitial spaces, and are capable of capturing macrofauna such as fully grown cumaceans and recently settled *Echinocardium* juveniles (Hartley, 1984).

5.8 Mussel beds

Where the sea-bed is particulate but has coarser material mixed in, there is potential for sessile organisms to become attached, and this leads to a considerable increase in the diversity of species present. The Common Mussel *Mytilus edulis* and the Horse-mussel *Modiolus modiolus* attach initially to small stones, but, as the colony grows, become attached to each other, so forming a relatively stable mat of animals over the surface of the sea-bed. Off the Isle of Man, for example, there is a belt of *M. modiolus* about 8 km long, and varying from about 250 m to 5 km in width, lying at a depth between 29 and 60 metres (Tebble, 1976). Off the Isle of Wight *Modiolus* is present at densities exceeding 1000 per square metre. Once these mussels become established, and even though they may be the dominant epifaunal species, their presence encourages other organisms to settle. In this way communities rich in species are established. In shallow areas of Sullom Voe, where the silt and sand is mixed with coarser material, there are dense epifaunal populations of *M. modiolus*, and many associated organisms. The shells of the mussel provide an extensive substratum for settlement of a wide variety of organisms, thus transforming a basically soft sea-bed into a hard one. One of the largest and most numerous at this site is the sea-squirt *Ascidiella*. The smaller *Ciona intestinalis* is also present as well as the red alga *Phyllophora* and various hydroids, bryozoans and serpulid polychaetes. Typical mobile animals are brittle-stars (*Ophiothrix fragilis, Ophiocomina nigra* and *Ophiopholis aculeata*), the Queen Scallop *Chlamys opercularis*, the whelk *Buccinum undatum*, the hermit crab *Pagurus bernhardus*, and the predatory starfish *Asterias rubens* and *Solaster endeca*. The mollusc and crustacean shells act as additional miniature habitats, providing a hard surface for attachment of small algae, hydroids, barnacles and so on. Similar communities occur at the eastern end of the English Channel where there are extensive beds of *Mytilus edulis*. However, species composition is slightly different because of

Fig. 5.6 The seagrass, *Zostera marina*, with epiphytic algae *Enteromorpha*, *Ectocarpus* and *Audouinella*. Associated animals are the Goldsinny *Ctenolabrus rupestris*, the squid *Sepiola atlantica*, and the hermit crab *Pagurus prideauxi*.

geographical factors. Neither *Ophiopholis aculeata* or *Solaster endeca* penetrate this far up the Channel, although the sunstar *Crossaster papposus* is present.

5.9 Seagrass beds

Few algae can maintain a hold on sediment bottoms unless there are small stones or shells present to which they can become attached, although species such as the Bootlace Weed *Chorda filum* may be present. *Zostera marina*, the commonest subtidal seagrass, is a flowering plant which is not fixed to hard surfaces but has roots that penetrate down into the sediment. It has strap-like, flexible leaves which reduce drag but still it will only flourish in areas stable enough for the roots to maintain a hold. Once established then the tangle of roots further increases stability and prevents plants from being torn away.

Water movement has a significant effect on community structure within seagrass meadows. In high current areas, or areas where there is strong tidal flushing, detached or decaying leaves are washed away, together with other detritus that may have been trapped between the stems. However, organic build-up will tend to be greater in areas where seagrasses are present, simply because current flow is reduced by the swards of leaves. The accumulated debris provides food for scavengers and deposit feeders which shelter on the sediment surface amongst the stems and leaves. Small wrasse and crustaceans, and the cephalopod *Sepiola* are amongst the mobile animals found in association with *Zostera*. The seagrass leaves, and especially the tip, often support epiphytic algae such as *Enteromorpha* and *Ectocarpus*, as well as hydroids and 'films' of ciliates and bacteria.

5.10 Maerl beds

Maerl consists of nodules of branched, calcareous red algae which lie unattached on the soft sea-bed in areas sheltered from heavy wave action but where there is sufficient tidal flow to prevent burial by fine sediments. They are also restricted to shallow, well-lit waters, and so are relatively sparsely distributed around the British Isles. One of the best known maerl beds is in the Fal Estuary, where the major species are *Phymatolithon calcareum* and *Lithothamniom corallioides* (Farnham and Bishop, 1985). The beds at this site, as in other areas such as Galway Bay and the Shetlands, support a diverse community of associated organisms. This is partly because of their open network which provides small niches for crustaceans such as the squat lobster *Galathea squamifera*, the Long-clawed Porcelain Crab *Pisidia longicornis*, Hairy Crab *Pilumnus hirtellus* and hermit crabs, *Pagurus* spp. Swimming crabs are often abundant on the surface of the maerl, and there are many sessile, epiphytic species, including hydroids, and a rich but seasonally variable assemblage of algae, many of which are red foliose and filamentous species. These are also found in other situations, but at least one of the crustose species, *Cruoria cruoriaeformis*, appears to be restricted to living maerl.

6

Development and Change in Benthic Communities

The preceding chapters have described some of the 'typical' communities associated with different areas of the sea-bed, and have perhaps given the impression that these communities are static entities which, once established, seldom change. In reality this is far from the truth, for their component species have to contend with environmental changes and biological interactions which may produce significant alterations in overall community structure. This chapter looks at the way that communities develop, the types of changes that occur, and some of the pressures that induce these changes.

6.1 Community development

6.1.1 Collecting the data

Studies on community development have concentrated on sessile benthos, because these communities are discrete and the biota static. The biota of open waters is not organized into communities in the same way because its components are constantly on the move, not only in relation to each other but also to space itself.

Sea-bed studies have often involved the use of artificially cleaned, hard surfaces. A common approach is to suspend settlement plates in the water, and study subsequent colonization, but this cannot be expected to mimic precisely the conditions and interactions on the sea-bed itself as the community develops. A more natural picture emerges from the study of discrete areas from which all sessile epibenthos has been removed, and this type of study also lends itself to manipulative experiments such as the exclusion of grazers or predators. Useful data have also been gathered by studying colonization on wrecks or other structures which make their appearance on the sea-bed from time to time, and are bare of all life at the outset. Some areas of the sea-bed are subjected to regular or unpredictable physical disturbances, with the loss of some or all of the biota, and recolonization has also been studied under these completely natural circumstances.

Sessile biota is renewed by settlement of larvae or spores which are very small, and need to be examined microscopically. Generally 'representative' settlement plates, small rock chips or whole stones are removed to the laboratory, but detailed examination has also been carried *in situ* by using a specially adapted incident light microscope (Kennelly and Underwood, 1984).

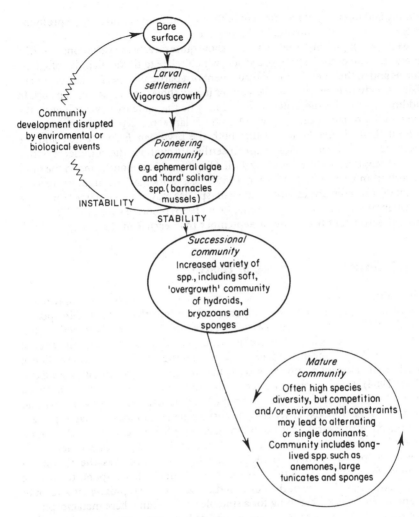

Fig. 6.1 Cycles of development and change in sessile hard-bottom communities.

6.1.2 Seasonal trends

An important factor influencing colonization by sessile biota is the time of year that the bare surface becomes available. Settlement on artificial substrata can of course be studied throughout the year, and the colonization patterns seen will reflect the availability and abundance of larvae during that time. The same principle is true for colonization of surfaces that have become bare as a result of natural processes. In the temperate waters of the British Isles there are likely to be more surfaces made bare during the winter as a result of seasonal die-back of algae, bryozoans, hydroids and other organisms, and the destruction of biota during winter storms. An essential point, however, is that variations in the timing and intensity of recruitment of the earliest colonists often have profound effects on the whole successional sequence.

Replicate experiments using panels show quite clearly the variations in settlement that can occur in time and space, which relate to the availability of young forms and to their behaviour. Recruitment of a particular species may occur at a fairly steady rate throughout the year, or it may show distinct seasonal peaks. In addition, although the general pattern may be the same from year to year, the availability of larvae often varies. A study of larval recruitment onto stones on Chesil Bank (Dorset) showed that both the bryozoan *Electra pilosa* and the serpulid worm *Pomatoceros lamarcki* settled throughout the summer months, even though the numbers settling varied from month to month, and in the same month from year to year (Warner, 1985). Less work has been done on recruitment of soft sediment infauna, but again species may show settlement patterns throughout the season, For example, Rees (1983) reported substantially greater bivalve settlement occurring in September than earlier in the year.

6.1.3 Settlement — chance or choice?

Undoubtedly a great deal of chance is involved, but it is also well known that many larvae have behavioural responses which lead them to settle in specific localities. For example, certain ascidian larvae are negatively phototactic, selecting shaded surfaces in preference to well-lit ones. The adaptive significance of this type of behaviour in ascidians has been investigated by looking at the fate of newly settled juveniles moved into shaded and unshaded locations (Young and Chia, 1984). Growth proved to be slower and mortalities higher in unshaded areas, due to siltation, overgrowth by filamentous algae (in shallow water), and grazing by gastropods. It seems that post-settlement mortality may represent the selective pressures that maintain negative phototaxis in the behavioural repertoire of ascidian larvae. Presumably the settling responses of other sessile species helps guide them to the microhabitat that provides the maximum chances for survival. It is not just sedentary animals that respond to specific cues in this way. Larvae of the dorid nudibranch *Adalaria proxima* is known to spend up to 2 weeks searching for a suitable substratum where metamorphosis can occur. In laboratory tests they settled only on the bryozoan *Electra pilosa* which is the preferred prey of adults. Chemosensory mechanisms are thought to be of paramount importance, and other nudibranchs including *Tritonia hombergi* are known to be similarly adept at locating prey (Thompson, 1976).

In addition to, or instead of, responding to the physico-chemical environment, larval settlement may also be influenced by biological factors, including intra- and inter-specific responses. For example larvae of the hydroids *Gymnangium montagui* and *Nemertesia antennina* prefer to settle at the bases of an existing colony, and in this way large clumps gradually develop. Species such as the bivalve *Mytilus edulis*, the ascidian *Molgula complanata* and the barnacle *Elminius modestus* show aggregated settlement, with swarms of larvae settling together. This might perhaps be due to swarms of larvae being present in the water at particular times, but in barnacles aggregated settlement is by choice rather than chance. The accepted view of gregarious settlement in

Semibalanus balanoides is that there is an interaction between adults and cyprids, involving recognition of adult cuticular proteins by the exploring larvae. Other factors such as surface contour, light intensity and water flow are also involved. Colonization of surfaces devoid of adults has in the past been attributed to indiscriminate settlement of old cyprids, or to sensitized cyprids that are more willing to settle anywhere after touching an adult. However, recent work shows that a pattern of gregarious settlement can occur simply as a result of larva–larva interactions (Yule and Walker, 1985). This is because exploring cyprids leave chemical 'footprints' in the form of antennular secretions which increase the rate of settlement of larvae. As more cyprids appear they reinforce their own and other footprints, and so the pattern of gregarious settlement develops. Another interesting adaptation these workers found was that the 'footprint' persisted on glass surfaces for at least 3 weeks, which was the average duration of the *S. balanoides* settlement season in the Menai Straits, where the experiments were conducted.

Animals colonizing soft substrata may also respond to specific cues. For example, organic matter and the by-products of organic matter such as hydrogen sulphide, act as a strong stimulus for the larval settlement of the polychaete *Capitella capitata*.

6.1.4 Successional sequences

A lapse of several days or weeks occurs after a clean settlement plate has been placed in the water before there are clearly visible signs of colonization. However, it is known that dissolved organic matter can become adsorbed onto the surface within hours, followed swiftly by a succession of bacteria. Within a few days there is a film of protozoa, diatoms, fungi and yeasts, as well as bacteria which provide food for larval colonists.

The successional sequences that follow depend to some extent on abiotic factors, but there are also biotic factors involved, many of which are interrelated. Growth rate and longevity of each species within the succession is important, but in addition some species may modify the habitat in such a way that others are able to invade. For example, it has been found experimentally that a canopy of the hydroid *Tubularia larynx* formed on fouling panels greatly enhanced settlement of the ascidians *Ciona intestinalis* and *Ascidiella*, which subsequently monopolized the entire substratum (Schmidt, 1983). Probably this enhanced recruitment was due to the shading and current-reducing effects of the hydroid canopy. A similar effect has been described for communities developing on North Sea oil platforms where currents over 300 cm sec^{-1} remove many of the fouling organisms but leave those such as barnacles that are firmly attached. The presence of barnacles can produce slack spots in which larvae of other species can settle, and thus even in a moderately fast current, a diverse community can develop. Conversely, there may be inhibition rather than facilitation, because what is advantageous for one species may be disadvantageous for another.

As the community develops so there is an increase in competitive interactions as individuals or colonies come into contact with each other, and there is also pressure from grazing and predation. Some of these points are illustrated by looking at the results of a study which compared colonization and succession at two sites. At one of these the primary colonists were colonial bryozoans, but these were then out-competed by solitary sea-squirts, which grew faster and to a relatively large size, so occluding the space beneath (Shin, 1981b). At a second site a more diverse community developed, with co-existence of many sedentary polychaete and colonial bryozoan species. It seems that in this case community development was being controlled more by abiotic factors than by interactions between species or individual animals. Free space remained on the plates, and it seems that the development of sessile communities may have been hindered by relatively slow current flow and heavy loads of deposited and suspended silt.

The significance and effects of grazing and predation on community development are still not entirely explained, partly because the experimental technique of caging alters the physical environment within the enclosure, as well as excluding predators and grazers. Schmidt and Warner (1984) took a closer look at the problem by reviewing some of the work already done and carrying out their own experiments to try and explain more clearly the relationship between caging and succession. They noted that many of the caging experiments showed an increase in the abundance of colonial and solitary ascidians, and a decrease in others, such as barnacles. Their own experiments mirrored these results, with caging causing a dramatic increase in the abundance of the solitary ascidians *Ciona intestinalis* and *Ascidiella aspersa*, while significantly reducing cover of the barnacle *Elminius modestus*. Previous workers generally attributed changes such as these to predation by fish or other animals, but Schmidt and Warner suspected that it could be the effects of the cage itself, rather than the exclusion of predators that was in many cases causing the faunal changes. In particular they thought that caging could be exerting its effects on the contained communities by reducing water movement and light intensity. This hypothesis was tested and verified by using a variety of control cages with windows of various sizes cut out, to separate the variables. The results showed that while settlement of barnacles and hydroids was reduced by reduced current speed, settlement of solitary ascidians was enhanced, and also occurred at lower light intensities. Predation was apparently playing only a marginal role in influencing community development because the communities on all types of caged panels, whether open or closed, were very similar, yet there was a marked difference between these and the uncaged controls. Much the same conclusion has been reached with regard to sediment communities, although again there has been some confusion over interpretation of the results of caging experiments (Gray, 1981). One fairly consistent feature of successional change amongst sessile hard-bottom biota is that primary colonists are often solitary forms, while secondary colonists are colonial. Goodman and Ralph (1981), studying the development of fouling communities on the Forties Platform in the North Sea, found a distinct progression over 2 or 3 years from a hard colonizing community of tubeworms, barnacles and saddle-oysters to a soft overgrowth community consisting of an extensive turf-like layer composed of large numbers of hydroids, encrusting bryozoans and sponges.

6.2 Stability versus change in established communities

Although relatively few long-term monitoring studies have been carried out, it appears that some benthic communities reach a mature and stable climax condition where species composition, abundance and biomass show only marginal changes from month to month and year to year. For example, when Hoare and Peattie (1979) resurveyed a transect across the Menai Straits after 20 years they found high biological stability, despite some population fluctuations. However, it is not uncommon for a climax community to be in a state of dynamic equilibrium, with some at least of the constituent species fluctuating in abundance but showing a recognizably consistent pattern when the community is studied over a period of years. One of the reasons for this is that in a crowded community there are constant interactions between the organisms as they overgrow each other, are eaten, or die-back at certain times of year, leaving spaces for others to colonize (Fig. 6.2). These types of fluctuations also occur amongst the mobile element of the community. For example certain fish and crustaceans migrate regularly from one area to another at certain times of year, so altering the appearance of the communities with which they are associated in a fairly consistent way.

Finally there are communities which are more drastically disrupted and may never have the opportunity to approach the climax condition. The degree and timing of disruption and change is determined by environmental and biological forces such as the physical stability of the substratum and the life history of the algal and animal species contributing to the community either as primary colonizers, or through their role as a substratum for other organisms. In the temperate climate of the British Isles certain areas are severely affected by seasonal environmental changes, and this usually causes annual or shorter term patterns of oscillating dominance. These periodic disturbances, which may range from the effects of storms on the one hand to development of oxyclines on the other, prevent climax communities from developing. In the same way, a transient community develops where there is a cycle of change due to predation or the short-lived nature of certain species. For example the hydroid *Nemertesia antennina*, which acts as a host for many associated species, dies back to a perennial holdfast every 4–5 months, which means that the communities associated with the stems inevitably are disrupted. A similar situation occurs with the kelp *Laminaria hyperborea* where mature communities develop on or in the holdfast, but not on the blade, which is shed annually (p. 72).

A climax community is more likely to develop in an area that is environmentally stable, and for the benthic fauna this applies both to soft and hard-bottom communities. Community change in soft substrata has been studied by regular sampling over a number of years, with analysis of biomass and species composition to detect change. This is more easily achieved than hard-bottom sampling, where there is often a higher degree of spatial variation.

A comparison of sandy-bottom areas of the sea-bed at opposite ends of the spectrum of physical regulation reveals some of the possible differences in community structure and dynamics. One recent study compared two sites off

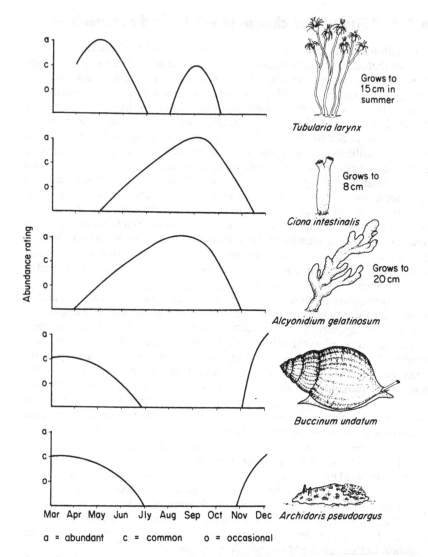

Fig. 6.2 Seasonal variation in abundance of three sessile and two mobile species on the submerged timbers of the 'Mary Rose' over the period 1980–1982. Adapted from Collins and Mallinson (1984).

the North Yorkshire coast; one inshore at a depth of 11 m, the other offshore at 45 m (Atkins, 1983). The offshore site was more physically stable, and the sediment poorly sorted. The community of polychaetes (54 spp.), bivalves (16 spp.), echinoderms (6 spp.) and amphipods (3 spp.) showed high seasonal stability, with no significant annual cycles in numbers of individuals, or numbers and diversity of species. In comparison with the inshore site, this site was both physically and biologically more stable. The dominant polychaete at this site, *Melinna cristata*, is typical of species associated with physically stable environments in that it has a conservative rather than an opportunistic life history strategy, living up to 5 years and producing non-planktonic larvae which last only about 3 days.

At the inshore site, which was strongly affected by storms, species diversity was much lower and there was re-establishment each year of a basically annual community by heavy settlement of all the dominant species at the end of the summer. This was followed by a winter decline, so that at the beginning of the spring the community was at a relatively stable level. Often a decline in numbers over this period can be attributed to winter storms, but in this case, a

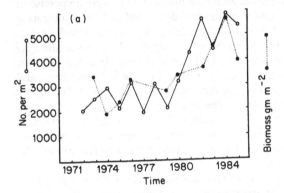

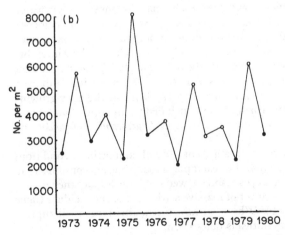

Fig. 6.3 Variation and stability in the benthic fauna of silty sand at a depth of 55 m off the Northumberland coast. From Buchanan and Moore (1986), and Buchanan *et al.* (1986). (**a**) Variation in the total number of individual animals per square metre (taken on a 0.5 mm sieve) and the biomass (ash-free dry weight). Sampling annually, in March, reveals a stable period (1974–1980), followed by a period of change. (**b**) Analysis of numbers per square metre during the stable period reveals an annual cycle of abundance due to summer recruitment, and a biennial cycle where a low March figure (•) is followed by a high September figure (○).

correlation was not proven, and it was suggested that the population was being regulated by density-dependent factors (Atkins, 1983). Whatever the regulatory mechanism, the pattern of rapid cycles of change in unpredictable soft-bottom habitats, and relative stability in undisturbed areas is well known. However, as Fig. 6.3 illustrates, even relatively undisturbed areas show cycles of annual and longer-term change.

The same sort of picture emerges for hard substrata, as illustrated in a study by Warner (1985) of two contrasting sites; one an old steel wreck, the other a bank of mobile stones. The community on the stones was a successional one dominated by rapidly-growing annual species. Disturbance and scouring during winter completely removed all sessile life, and there were no persistent individuals. Despite this, a similar community developed each year, indicating that the same array of larval species was available for settlement each spring. Two of the most successful colonists were the serpulid *Pomatoceros lamarcki* and the bryozoan *Electra pilosa* (Fig. 4.8). A community of algae developed on the top of the stones, but even so the stones never became fully colonized, and so competition for space did not play any significant part in structuring the community. A characteristic of successional communities, whether their development is arrested by physical or biological factors, is that they contain a high proportion of opportunistic species, which are capable of rapid colonization and growth, even though they may have a short life expectancy or may be reduced to a resting phase each year.

In contrast the community on the wreck was seasonally cyclic, but otherwise appeared to be stable, and was assumed to have reached a climax. It contained slow-growing species, individuals of which persisted from year to year, and faster-growing species which reappeared each year. Amongst the persistent species were the sponges *Myxilla rosacea* and *Dysidea fragilis*, the octocoral *Alcyonium digitatum*, and the tube-dwelling amphipod *Corophium sextoni*. Some of the annually reappearing species simply died back over the winter, but maintained a presence. For example both the bryozoan *Bugula plumosa* and the solitary ascidian *Clavelina lepadiformis* spread rapidly in the spring from over-wintering stolons, often temporarily over-topping some of the slow-growing species. Other annually reappearing species included the hydroids *Eudendrium ramosum* and *Obelia dichotoma* and the sponges *Leucoselenia botryoides* and *Scypha ciliata*.

Even though a site may be in an apparent state of equilibrium, conditions may arise to cause destabilization. For example, a rocky-bottom community in Norway was studied for several years and showed stable cycles, but then underwent a complete change as a result of massive settlement of the ascidian *Ciona intestinalis* (Gulliksen, 1980). Why, and how often, this type of disruption occurs in established communities is not known.

It is not only communities as a whole that show seasonal cycles of change; some of the long-lived plants and animals also do so. This is especially noticeable in kelp plants, which annually lose and regrow their blade, and also in various hydroids and ascidians which die back to persistent stolon-like structures from which new colonies or individuals arise the following year. Other species show distinct although less drastic changes. For example, individual colonies of the soft coral *Alcyonium digitatum* studied by Griffiths and Dennis

(1984) showed a well-defined cycle of activity. In October and November the colonies became largely inactive and shrank to their smallest dimensions, becoming leathery and discoloured. In late December a few colonies were expanded and feeding, but this activity increased until a peak was reached in May/June, by which time the colonies had expanded to about 4 times their November size. The activity then waned gradually over the summer. In contrast, *A. glomeratum* showed no annual cycle.

6.3 Effects of grazing and predation on community structure

Grazers and predators are consumers, and so inevitably have an impact on the distribution and abundance of their food supply. Often this is not immediately obvious because a state of dynamic equilibrium exists, whereby the food is being replaced at about the same rate as it is consumed, with the consumer maintaining a steady population linked to this renewal rate. In other cases the populations of predator and prey or grazer and sessile biota follow each other in rather more dramatic cycles, sometimes linked to seasonal changes in abundance of one species or another. For example, spring 'blooms' of the solitary hydroid *Tubularia* are intensively grazed by nudibranch molluscs, just as phytoplankton blooms are subject to the attention of zooplankton.

An important all-year-round grazer on hard-bottoms is the sea-urchin *Echinus esculentus* (Fig. 6.4), whose activities have wide ecological effects,

Fig. 6.4   The sea-urchin, *Echinus esculentus*.

especially within the kelp forest, where they influence the distribution and abundance of kelp plants by grazing both on the sporlings and the mature plants. The extent to which *E. esculentus* selects its food is not known, but some preferences have been discovered, and illustrate that it is not only the grazing but also the selectivity that is important. For example, laboratory Y-maze experiments have shown that *E. esculentus* from a kelp forest supporting *L. saccharina* and *L. digitata*, when offered a choice between these algae, showed a clear order of preference, with encrusted *L. saccharina* most preferred, followed by encrusted *L. digitata*, clean *L. saccharina* and finally clean *L. digitata* (Bonsdorff and Vahl, 1982). It was suggested that, at least at this site, selective grazing of *L. saccharina*, which is an opportunistic species, was giving *L. digitata* a competitive advantage and enabling it to extend its range to deeper water.

The impact of grazing by *E. esculentus* appears to be particularly critical at the lower edge of the kelp forest, where conditions for growth of algae become marginal. Removal of *Echinus* from this zone has been found to cause a significant downward extension of the forest within 2 to 3 years (Jones and Kain, 1967). Conversely, in a Californian kelp forest, heavy fishing of lobsters, which prey on *Echinus*, is thought to have caused an increase in urchin populations and contributed to episodes of destructive urchin grazing observed since the 1950s (Tegner and Levin, 1983).

A rather smaller herbivorous grazer, but one which still has an important ecological impact, is the Blue-rayed Limpet *Patina pellucida*, which commonly grazes on kelp blades, but often migrates down to the stipe and holdfast. At the base of the stipe it hollows out cavities by its activities and these weaken the plant and make it more likely to collapse or be torn off the sea-bed.

Another species that can have a considerable impact is the starfish *Asterias rubens*, which aggregates in high densities on mussel beds, methodically attacking shell after shell. In this way, entire mussel beds may be decimated, only to be built up again as a result of larval recruitment. Starfish have even been employed as a means of controlling mussels fouling the legs of oil rigs, although this first necessitated a path to be cleared through a mass of *Metridium senile* so that the starfish could migrate up from the bottom of the legs to the mussels at the top (Ralph and Goodman, 1979). The protective value of *Metridium senile* has also been demonstrated in the laboratory where it was found that the presence of the anemone on the valves of *Mytilus edulis* significantly reduced predation by *Asterias* (Kaplan, 1984).

The opportunities for grazing in soft-bottoms is naturally extremely limited by the lack of sessile fauna and flora. However, it is known that a form of grazing can occur where flatfish crop the tips of bivalve siphons and the palps of polychaetes. These are then regenerated in much the same way as many of the hard-bottom animals which are able to regrow from remnants.

7

Behavioural Interactions on the Sea-bed

7.1 Resource partitioning

One reason why the sea, and particularly the sea-bed, supports such a dense array of life is that the essential resources of space and food are, to some extent, partitioned out, not only between species, but often between different stages in the life cycle of the same species. For example, *Mytilus edulis* has an interesting strategy which avoids competition between established adults and newly settling young. Initial settlement takes place primarily on filamentous substrata, away from established mussel beds. However, attachment is only temporary, and the larvae can break away from their byssus attachments and explore locally or migrate further afield by re-entering the plankton. They are aided at this stage by their long byssal threads, which are used as flotation devices (Lane *et al.*, 1982). This type of strategy may be clearly apparent in other species, as when two species live next to each other but feed on quite different food sources, or feed on the same food source but are spatially separated. Some species appear to have precisely the same requirements, but close examination of where and how they live often reveals small but significant differences which presumably help to reduce competition and so increase the likelihood of that species surviving. For example, a study of detrivores feeding at the sediment/water interface on a muddy sand bottom showed that of the three major taxonomic groups polychaetes mostly used palps and tentacles to collect food from the sediment surface and from suspension, while molluscs filtered out suspended material from just over the sea-bed, and small crustaceans stirred up the surface layer and filtered the resulting suspension (Eagle and Hardiman, 1976). In another study resource partitioning was shown to exist in some of the suspension feeders occurring in kelp beds. The filtering efficiency of three species of bivalves, an ascidian and a sponge was compared, and it was found that while the bivalves retained the largest particle most efficiently, the sponge showed highest retention for the smallest particle, and the ascidian was most efficient in the medium to large size range (Stuart and Klumpp, 1984).

7.2 Competition

Inevitably, even though there is partitioning, there is bound to be competition for space and food, especially on the sea-bed where many of the organisms are sedentary and unable to migrate in order to exploit the resources of another area, or to avoid the attention of grazing and predatory animals. In open water, space is not at a premium, and organisms are constantly on the move, so interference is less of a problem.

Benthic organisms interfere and compete with each other by both subtle and openly aggressive means. On hard substrata shading, overgrowth, undercutting and abrasion all come into play. These competitive mechanisms may occur as a result of the life style and growth characteristics of the organism involved, for example the adoption of particular growth forms to prevent smothering (Fig. 7.1), or be achieved by deliberate aggressive behaviour. In some cases where there is competition for the same food resource, one species noticeably predominates. This has been shown to be the case in a dense bed of the brittle-star *Ophiothrix fragilis* (George and Warwick, 1985). Here, although a diverse assemblage of other species was present, they were characterized by a slow growth rate and a small size, and production was completely dominated by *Ophiothrix fragilis*. It was suggested that the slow growth rates and small size of the other animals may have been due principally to the fact that *O. fragilis* monopolizes the suspended food resource with an umbrella of feeding arms.

Competition through shading affects the growth and distribution both of algae and of animals containing zooxanthellae. The kelp, *Laminaria hyperborea*, with its rigid stipe, produces a shading effect by its canopy of fronds, so has a competitive advantage over species such as *L. digitata* which has a flexible stipe and cannot lift the fronds so high. It has been shown that if patches are cleared within a forest of *L. hyperborea*, the unshaded ground is colonized ini-

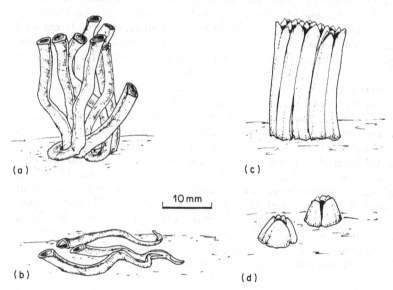

(a)

(c)

10 mm

(b)

(d)

Fig. 7.1   (**a**) Perpendicular growth form in *Pomatoceros triqueter* in response to competition for space on the Beatrice oil platform in the North Sea (drawn from a photograph in Forteath *et al.*, 1983); (**b**) 'normal' growth form. (**c**) Columnar growth in *Balanus crenatus* from the same location. Hummocks formed from these growths are long-lasting in calm conditions; (**d**) 'normal' growth form.

tially by *L. digitata*, but that these plants are then gradually eliminated as the *L. hyperborea* canopy is re-established (Kain, 1979). In the normal way, natural death of a mature plant of *L. hyperborea* results in rapid upgrowth of young plants from the understorey to replace the canopy. *L. hyperborea* also has a growth strategy which gives it a competitive advantage over other species. The growth of the fronds in this species is at its peak in May, and ceases in late summer. Part of the organic matter assimilated by the fronds is stored, and then is used in early spring to form new fronds, before light levels are sufficiently high to enable growth to take place (Luning, 1971). In contrast, *L. digitata* grows prolifically from spring to late autumn, but does not store organic material, with the result that it has a slower start to growth the following year.

Abrasion of benthic biota by stones and sand is well known, but abrasion by both algae and animals also occurs, and can be seen as a form of competition because it often results in an exclusion zone being set up around the organism concerned. Algae such as kelps with their tough, whippy fronds are particularly successful, but other species may have a similar effect on a smaller scale. Some organisms achieve the same effect by having a mechanism for keeping their immediate patch of space clear. Best known are the sweeper tentacles of anemones (section 7.3).

In soft-bottom habitats there is little sessile epibenthos, and thus competitive mechanisms such as shading and overgrowth which exist amongst the sessile biota of hard-bottoms are unlikely to be important. However, other types of biological interference come into play, for example between the sedentary tubiculous animals on the one hand and the truly mobile infaunal element of the fauna on the other. Where tubes occur then the freedom of movement of the larger burrowing animals such as polychaetes, bivalves, crustaceans and echinoids is curtailed and their populations restricted. The roots of seagrasses apparently act in a similar way (Brenchley, 1982). Clearly this interference competition works both ways, for example, a well-established population of deposit feeders which rework the sediment will inhibit the activities of suspension feeders.

7.3 Defensive and offensive mechanisms

Pressure from competition, grazing and predation, combined in predators with the need to capture prey, has resulted in the evolution of a wide range of defensive and offensive mechanisms amongst benthic organisms. The organisms most in need of defensive mechanisms are soft-bodied sessile organisms such as anemones. These have stinging cells (nematocysts) on the tentacles and some have mesenteric filaments with thread-like acontia which are also heavily armed. These can be ejected through the mouth of the anemone, or through specialized pores in the body wall and they play an important role in defence and intraspecific confrontations, as well as in subduing prey. However, even the defences of acontian anemones can be penetrated; for example the Tompot Blenny *Parablennius gattorugine* is reported to feed on *Sagartia* (Milton, 1983), and the sea-slug *Aeolidia papillosa* on *Cereus*, *Sagartiogeton* and *Metridium*. When given a choice, however, *Aeolidia* has a clear preference for various actinian species, and avoids *Metridium senile*.

Echinoderms are also highly successful at defending themselves. For example, echinoids not only have spines which can be directed towards predators, but also mobilize their globiferous pedicellariae which contain poison and can be pinched onto the flesh of potential predators, such as large starfish. However, certain animals can still penetrate these defences, for example mature Haddock are known to feed on *Echinus acutus*.

Many animals rely primarily on concealment rather than possession of defensive structures. Visual camouflage is well known, and there are numerous examples where combinations of colour, texture, shape and movement are involved. Amongst the masters of camouflage are cuttlefish, with their ability for rapid change designed to confuse potential prey. Benthic fish such as scorpionfish, anglerfish and flatfish match themselves against the background, and often are extremely difficult to detect if they remain still. Numerous smaller animals are equally well adapted, including caprellids, small crabs and various nudibranchs (Fig. 7.2).

Wicksten (1983) points out that visual camouflage will be of no use against predators that lack eyes or hunt in the dark. Echinoderms, nudibranchs and certain gastropod molluscs for example, hunt by chemosensation and touch, tracking their prey along a sensory gradient. Clearly in these cases it would be advantageous for the prey species to have some means of chemical or tactile camouflage. Wicksten suggests that encrusting animals such as sponges might fulfil this role by producing chemical products (cryptic odours) that in some way mask the metabolic by-products of the prey organism beneath. Presumably encrusting organisms also disguise the usual 'feel' of the organism beneath, thereby further confusing the predator. This adds another dimension to the

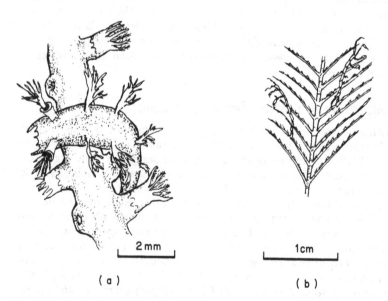

(a)   (b)

Fig. 7.2 Camouflaged animals. (**a**) The nudibranch *Tritonia odhneri* on *Eunicella verrucosa*. (**b**) Caprellids on the hydroid *Nemertesia*.

adaptive significance of some of the commensal relationships that occur between organisms on the sea-bed (p. 72).

A different type of chemical defence strategy involves the use of toxins which render the prey species distasteful or poisonous. These toxins are secondary metabolites which the animal accumulates in its body, and they have been identified from a range of sessile species, as well as from nudibranchs, echinoderms and fishes. Sessile organisms may also use these biologically active compounds as a form of antifouling to prevent the larvae of other species from settling and growing on their surfaces, and also in competitive interactions with established neighbours. Hydroids, for example, have been shown to be sensitive to chemical substances produced by certain brown algae which act as antifoulants (Sieburth and Conover, 1965).

7.4 Reproductive and life history strategies

Many benthic organisms produce planktonic young; a strategy which helps to spread the species by providing an opportunity for colonization of new areas. It is also possible that the pelagic environment offers a better supply of food for the rapidly growing young. On the other hand there is an energy cost to pay because many larvae will be carried away to unsuitable areas.

There are various sessile organisms that spread by some form of asexual reproduction. For example, the Plumose Anemone, *Metridium senile*, can reproduce asexually by longitudinal fission or pedal laceration, and in doing so, forms clones of genetically identical individuals. It has been found that clonemates become aggressive against neighbouring clones, but that aggression is only towards members of the same sex. This results in males coming closer to females and vice versa, so presumably increasing the chances of successful fertilization (Kaplan, 1983).

Some species cope with a high level of predation or grazing by having an extremely fast growth rate and turnover to replace losses. Others, which are generally slower growing, and can often recover from being partially eaten or grazed, put their energies instead into defensive mechanisms. The first group are often referred to as r-strategists, so called because of their high reproductive rate and rapid growth. These are the 'boom and bust' opportunistic species able, when conditions are favourable, to respond quickly and build up large, even if transient, populations. Barnacles, the keel-worm *Pomatoceros* and the hydroid *Tubularia* are all typical hard-bottom opportunists, readily colonizing spaces on the sea-bed as they become available. *Capitella* species show a similar strategy in soft substrata. Despite their prodigious growth rates, a proportion of these opportunistic settlers may be out-competed or disrupted in other ways before they have time to reproduce.

On the other hand are the K-strategists, with K signifying the carrying capacity of the environment. These species tend to live in more predictable and stable environments, and put their efforts into slow growth and long life. However, not all organisms fit neatly into these categories. For example, *Mytilus edulis* is opportunistic in its production of large numbers of larvae which often proceed to blanket extensive areas of available substrata. Yet they are long-lived species which can adapt to low food levels by growing slowly, in this way fitting the K-strategy (Gray, 1981).

Another interesting link is that between the type of development and duration of pelagic larval life, and the degree of fluctuation shown by populations of benthic invertebrates. Populations that do not fluctuate widely from year to year are commonly those that have a short pelagic life or a non-pelagic development, whereas fluctuating populations tend to be those which have larvae with a long planktonic life (McCall, 1978).

7.5 Migrations

Migratory movements amongst benthic crustaceans are well know, especially amongst the commercial species, and in some cases are known to be tied to the breeding cycle. For example, in the English Channel mature female Edible Crabs migrate from eastern areas and congregate in the autumn in offshore areas off south Devon, at which time they are ready to spawn. It has been suggested that this may be necessary to ensure a suitable substratum for the berried crabs, as well as allowing for distribution of larvae on the easterly flowing current in the Channel (Bennett and Brown, 1983). Females need a soft substratum of sand or gravel for spawning because they need to be able to lower the abdomen into it in order to ensure attachment of eggs to the pleopods. Both males and females also show nomadic movements, both along the Channel and from deep to shallow water.

Less is known about the Crawfish *Palinurus elephas*, a south-western species which is evidently recruited from outside the area, since juveniles are not represented in the population. Little is known about how they reach the area, but fishermen have reported seeing shoals of Crawfish swimming at the surface, with their tails spread out and antennae held clear of the water (Hepper, 1977). Related species from other parts of the world are known to march in long lines across the sea-bed. Several benthic fish species make regular migrations to inshore waters to deposit their eggs, and species such as the Salmon, Sea Trout and Sea Lamprey run up rivers to breed in freshwaters.

7.6 Intraspecific relationships

There are many ways in which individuals of the same species interact, but perhaps two of the most obvious are aggregating behaviour on the one hand and territorial behaviour on the other. Many sessile species occur in aggregations, instead of being randomly distributed over the sea-bed. This may result from larval distribution or behaviour, or through the successful competitive activities of the established organism. For example the anemone *Corynactis viridis* spreads by asexual reproduction, and often carpets extensive areas of rocky substrata. There are also mobile species that spend most of their time in aggregations. The brittle-star *Ophiothrix fragilis* is one such species, and in this instance the act of aggregating leads to increased food availability, which is rather the reverse to what might be expected. The reason for this is that a mass of brittle-stars with their arms interlinked is more stable than an individual animal, and so the mat of animals can become established in areas of stronger water flow which carries correspondingly more suspended material across the extended arms (Warner, 1971). Warner (1979) also suggests that, by interlocking arms and propping each other up, the animals have more 'free' arms for feeding. In addition the dense forest of extended arms may slow down the passing current and so lead to increased deposition of potential food particles from suspension.

These aggregations are apparently maintained by social interactions. When brittle-stars were separated from the mass and placed on the sea-bed nearby they immediately started walking until they encountered a mass of conspecifics. This achieved, they climbed into the aggregation and began feeding (Broom, 1975). Aggregating in this way, as with shoaling in fish, is thought also to provide protection against predators, and to increase chances of successful fertilization. Other species form aggregations only during the breeding season. One of the most spectacular is the spider crab *Maja squinado*, which masses into heaps, with the newly-moulted, soft-shelled females in the middle, and the hard-shelled males on top.

Very little is known about territorial behaviour in benthic species occurring in the British Isles, but there is little doubt that certain crustaceans and fish hold territories at some stage during their lives. The prime reason why an animal establishes and defends a territory is so that it can maintain an exclusive hold on certain resources, often in connection with feeding and breeding. Male Corkwing Wrasse, *Crenilabrus melops*, establish a breeding territory in which the eggs are deposited and protected. This area is defended vigorously, and the intrusion of all potential egg predators is prevented. Breeding territories are set up seasonally, and operate only for as long as they are required. Other animals have permanent territories which function primarily to protect a food resource, but also serve a useful purpose during the breeding season. It is probable that the Norway Lobster *Nephrops norvegicus* and the crabs *Inachus* and *Hyas* may be territorial, but much remains to be learnt from *in situ* studies.

7.7 Interspecific relationships

7.7.1 Living together

The sea-bed is a crowded place to live, and there are innumerable examples of species living together in some way without doing apparent harm to each other. By far the most common type of association is a casual one in which organisms benefit by finding a perch or niche for themselves on or in other organisms. When the 'perching' species, which may be either mobile or sessile, can also live independently then it is referred to as a facultative commensal. In contrast, an obligate commensal has an absolute dependence on its host and is never found living independently. Finally, there are the symbiotic relationships which are often obligatory for at least one of the species involved, and always benefit both parties.

7.7.2 Facultative commensals

Facultative commensals include all the sessile and mobile organisms growing or living as epiphytes on algae or epizooites on animals, and sometimes there can even be more of these commensals than there are organisms attached to the sea-bed itself. The most heavily colonized hosts are erect species such as kelps and medium- or large-sized algae, hydroids, and tunicates, but many other smaller- or lower-growing species also play host, as do mobile species such as crabs, whose hard carapace is well suited for sessile epizooites. Some of the benefits of being an epiphyte or epizooite can be deduced by looking at a few hosts and at the distribution of associated organisms. Sometimes species that are abundant on particular hosts may be rare or absent from the surrounding area, and this indicates how in these circumstances the host is playing an important role in providing a completely different habitat from that of the sea-bed itself.

Laminaria hyperborea

L. hyperborea is the most important kelp species around the British Isles and supports a wide range of epiphytes, whose distribution and abundance relates to the age and longevity of different parts of the plant, the gradient of abiotic conditions that occur from base to tip, and the structure of the plant itself. At the base of each plant is a sturdy holdfast consisting of a bunch of branching haptera which are attached to the substratum by their tips. The holdfast is a perennial part of the plant and new haptera are added each year so that a complex structure is built up, providing small holes and crevices where animals such as brittle-stars and small crabs can nestle or take shelter, and where there are surfaces for attachment of sessile biota. As many as 389 species have been identified from 72 holdfasts (Moore, 1973). Analysis of kelp holdfast communities has become an almost standard part of detailed survey work, especially since a link has been found between the type of community and the degree of pollution. Jones (1971) found that holdfasts from polluted waters close to urban development on the north-east coast of England contained an average of 44 individuals belonging to 8 species, while holdfasts of a similar size from unpolluted areas supported an average of 80 individuals belonging to 19 species. Turbidity arising from natural rather than man-made causes is also known to reduce species diversity of kelp holdfast fauna.

Many of the holdfast associates are suspension feeders. Barnacles, the bivalve *Hiatella arctica*, and the brittle-stars *Ophiothrix fragilis* and *Amphipholis squamata* are common. However, a particular feature of the holdfast is that it has a sediment-trapping effect, so favouring colonization by detrivores and scavengers such as nematodes, amphipods and small crabs. Tubiculous species of amphipod also use sediment to construct their tubes.

The stipe of *Laminaria hyperborea* is held up from the sea-bed, and has a rough surface which enables larval colonists to gain a foothold. It provides a silt-free habitat and an excellent vantage point for suspension feeders, and is generally encrusted with a mixture of sponges, colonial ascidians, bryozoans, barnacles and hydroids. Red algae such as *Palmaria palmata* and *Ptilota*

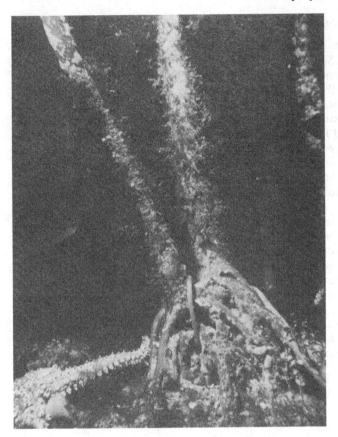

Fig. 7.3 Holdfast and stipe of *Laminaria hyperborea*, with epiphytic growths.

plumosa are also conspicuous colonists. Grazers and carnivores ascend the stipes in order to feed on the sessile organisms, although the sea-urchin, *Echinus esculentus* is liable to be knocked off onto the sea-bed in turbulent conditions.

The kelp blade supports a more restricted epibiotic community because it is short-lived, lasting only over the spring and summer months, and because it is lashed by the water and by blades of adjacent plants. However, it is often extensively covered by encrusting bryozoans such as *Membranipora membranacea* and the hydroid *Obelia*. These typically establish themselves in spring on the youngest parts of the blade and grow with it, rather than settling on the older parts which will be the first to be discarded. The sessile animals attract the attention of carnivorous nudibranchs, including *Polycera quadrilineata* and *Dendronotus frondosus*. There are also grazers such as *Gibbula cineraria* and *Patina pellucida* that feed both on attached algal 'films' and on the tissues of the plant itself. The blades of other kelp species support a similar assemblage of epiphytes.

Saccorhiza polyschides

This kelp has a bulbous holdfast which provides a microhabitat different to that of other species. When mature, the holdfast consists of a series of interconnecting chambers, and in addition, there are knobbly haptera which grow down from the bulb to the substratum. A total of 77 species have been recorded from a survey of holdfasts off the coast of western Scotland, with the greatest number of species occurring between the rock surface and the base of the bulb, where both shelter and food (detritus) are provided (McKenzie and Moore, 1981). Brittle-stars and polychaetes were present, but the fauna was dominated by the amphipods *Amphithoe rubricata* and *A. gammaroides*. Inside the bulb were large polychaetes, including *Nereis pelagica*, crustaceans such as *Galathea strigosa* and *Pisidia longicornis*, and various fish. The Gunnel *Pholis gunnellus*, Cling-fish *Apletodon microcephalus*, Two-spot Goby *Gobgiusculus flavescens*, Montagu's Sea-snail *Liparis montagui*, and juvenile Shore Rockling *Gaidropsarus mediterraneus* and Goldsinny *Ctenolabrus rupestris* have all been recorded from within the holdfast, and it seems that their presence may be the cause of the low density of sessile species on the inner walls of the bulb. They gain access to these inner cavities through holes in the bulb, especially on the base, and it seems that the amphipods resident in the crevices beneath the bulb are an important source of food for the fish. Male Clingfish and Two-spot Gobies both defend a nest which typically is established on the sides or roof of the inside of the bulb (Gordon, 1983).

Fewer species are associated with the outside of the bulb, and most of these

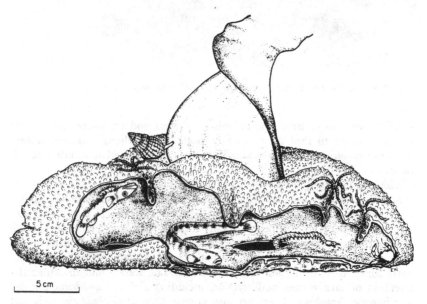

5 cm

Fig. 7.4 Fauna associated with the holdfast of the kelp *Saccorhiza polyschides*. On upper surface: *Nassarius incrassatus*, *Ophiopholis aculeata*. Within bulb: *Apletodon microcephalus*, *Pholis gunnellus*, *Nereis pelagica*, *Pisidia longicarnis*. Below bulb: amphipods.

are suspension feeders (e.g. *Membranipora membranacea*) or mobile herbivores and scavengers such as *Nassarius incrassatus*. Where crevices or invaginations occur on the upper surface, for example between bulbs, silt is trapped, and a suitable microhabitat formed for various species, including *Ophiopholis aculeata* and other brittle-stars.

Halichondria panicea

A diverse fauna has been found in association with this sponge; in one study 50 species were recorded from 41 samples (Peattie and Hoare, 1981). *H. panicea* often occurs in areas where there are strong currents, and clearly one reason why it is colonized by other species is that it provides shelter-spots in an area otherwise difficult to colonize. Most of the colonists are also found in association with other hosts, such as *Nemertesia* and *Laminaria*, although Peattie and Hoare found that the small crustacean *Caprella linearis* was present in very high densities, and thought that it was probably chemically attracted to the sponge. These suspension-feeding caprellids clung to the exposed parts of the sponge even when the current was flowing strongly, but most of the other species remained within crevices, only emerging when the water flow slackened. A complete food web was found, which could not be maintained in the absence of the sponge, where the rocky surfaces did not provide sufficient shelter from the strong currents.

Nemertesia antennina

These hydroids occur as slender stems about 30 cm high, sometimes alone, but more often in clumps. As many as 150 species have been found in association with a population of *Nemertesia*, and examination of single stems reveals clear distributional patterns which correlate with the requirements of each epizoic species, and the range of abiotic conditions from the top to the base of the host (Hughes, 1975). The epifauna was found to be dominated by suspension feeders such as sponges, hydroids, bryozoans, amphipods and ascidians. Deposit-feeding prosobranchs and predatory polychaetes were also common, although predators as a group numbered less than 4% of the epizooites. Passive suspension feeders such as the amphipod *Caprella* and the hydroid *Plumularia* live on the distal parts of the host so that they are elevated into a position where water movement is greatest and suspended food is continually passing by. In contrast, active suspension feeders are found lower down in slower moving water and deposit-feeding epizooites are almost entirely restricted to the holdfast. The distributional patterns are related primarily to gradients in water flow. The tidal current is slowed down both by the frictional drag of the sea-bed and the hydroid itself, so that the flow is considerably less at the holdfast end, and debris accumulates.

Pentapora foliacea

This erect, leafy bryozoan may live for 10 years or more and often supports a large number of epifaunal species. A study of *P. foliacea* from a depth of

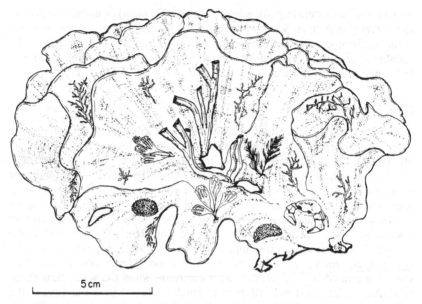

Fig. 7.5 Fauna associated with the bryozoan *Pentapora foliacea*, including the tube-worm, *Pomatoceros*, the crab, *Pisidia*, and the bryozoans, *Celleporaria pumicosa*, *Bugula* spp., *Scrupocellaria* spp. and *Crisia* spp.

15–20 m in strong tidal flow off the Welsh coast revealed that bryozoans were especially common amongst the sessile forms, and that the mobile epifauna was dominated by caprellids and the small crab *Pisidia longicornis* (Stebbing, 1971). The encrusting bryozoan *Bugula flabellata* occurred widely, and had stolons buried within the calcareous fronds of its host. It bores in a similar way into other calcareous or gelatinous bryozoans, but on rocks and shells the stolons simply ramify over the surface. Growth of the bryozoan *Scrupocellaria reptans* and the hydroid *Sertularia rugosa* was orientated towards the distal growing edge of the fronds, presumably elevating the colonies into a good feeding position. Bivalves such as *Musculus discors* and *Hiatella arctica* nestled in surface irregularities and in crevices between the fronds.

7.7.3 Obligate commensals

With these species there is an absolute dependence by one partner on the other, although in some cases the precise relationship is not fully understood. For example, the barnacle *Boscia anglicum* is an obligate commensal on corals and in south-west Britain is associated primarily with *Caryophyllia smithii*. *Boscia* retains the generalized barnacle apparatus for capturing planktonic food, and does not become deeply embedded in the coral skeleton. However, tissue of the coral host covers the surface of the barnacle except for the orifice, and death of the barnacle usually follows after withdrawal of this tissue, raising the possibility of nutritional dependence (Anderson, 1978).

7.7.4 Symbiosis

A well known hermit crab symbiosis is *Adamsia carciniopados* which lives almost exclusively on *Pagurus prideauxi*. If the two animals are separated the crab rarely survives for long, and the condition of the anemone eventually deteriorates (Manuel, 1981). *Calliactis parasitica* is usually found in association with *Pagurus bernhardus*, but may occur on other crabs, or on stones. *C. parasitica* actively transfers itself onto the crab, unlike the situation in the Mediterranean, where it is picked up by the crab *Dardanus*. Both *Adamsia* and *Calliactis* possess acontia, and these are readily emitted when the host crab becomes agitated (Manuel, 1981). Presumably both the anemones give some form of protection to their host. The Plumose Anemone *Metridium senile* may perform a similar function when it attaches to species such as *Modiolus* or *Mytilus edulis* on mussel beds. Laboratory experiments have shown that the presence of the anemone significantly reduces predation by the starfish *Asterias forbseii* (Kaplan, 1984). The anemones are thought to benefit from the feeding activities of their host, which circulates food in their direction.

8

Open Water Life

8.1 Introduction

Preceding chapters have been concerned solely with benthic communities, where the animals and plants are confined to the sea-bed; a thin crust in comparison with the wide open spaces of the pelagic zone. Many of the sea-bed organisms are buried or attached, some creep, crawl or burrow, but relatively few swim. Open water communities are quite different because space is virtually unlimited, yet the water provides no shelter, and only minimal support. Colonists must therefore be able either to float or swim in order to remain part of the pelagic community. Passive drifters that are able to direct their movements only within a very limited sphere (if at all), are known collectively as plankton. Mobile animals capable of swimming against the flow of water comprise the nekton.

Water is a much less complex and variable medium in which to live than the sea-bed, and for this reason open water life does not fall into such clearly defined communities. Another factor to consider is that, in open water, individual plants and animals are constantly on the move, not only in relation to each other, but also to space itself. However, this does not mean that open waters are uniform in terms of the diversity and abundance of species they support. The composition of the plankton is influenced to a considerable extent by water quality and current regimes, while the nekton responds not only to the type and abundance of plankton, but also, in some cases, to the type of sea-bed or benthic community. These interactions between nekton and sea-bed usually relate to feeding or breeding requirements.

Organisms living in open water are like their benthic counterparts in that they undoubtedly compete for food. However, unlike the situation on the sea-bed, there is probably little or no competition for space. Thus it is not surprising to find that pelagic organisms have a range of adaptations and devices which make them either more effective as predators or less easily taken for food, or both.

Most planktonic organisms are too small to be studied *in situ*, and there are many practical problems involved in making *in situ* observations on the nekton. This is because all the animals involved are effective swimmers with a well developed sensory system, and they tend to retreat rapidly in the face of intrusion. However, some information has been gleaned from lucky encounters, and by using video and other photographic techniques. Sea mammals always excite interest, and some details of their behaviour have been learnt by fitting them with tracking devices so that their movements can be followed.

Soft-bodied and gelatinous zooplankton organisms such as ctenophores, siphonophores, doliolids and some medusae, are extremely delicate and almost invariably are damaged when collected in nets, often emerging at the surface as a shapeless lump of jelly. In recent years considerable advances have been made in understanding the morphology and behaviour of these animals through the use of photography. Cameras operated both by divers and by submersibles have been used successfully (Madin, 1985). Non-living components of the plankton have been studied in the same way, including large detrital aggregations known as 'marine snow' which, on collection, look nothing more than a small blob of debris.

8.2 The plankton

8.2.1 Introduction

Suspended in the water column are teeming millions of free-floating plants (phytoplankton), animals (zooplankton) and micro-organisms; many so small as to be invisible to the naked eye. Table 8.1 gives an indication of the sizes involved. Some of the smallest planktonic organisms (e.g. bacteria and protozoans), are concentrated in a microscopically thin layer at the surface of the water where concentrations of dissolved organic matter (DOM) and inorganic salts are high. This specialist community is known as the neuston.

Common amongst the phytoplankton are diatoms, dinoflagellates and blue-green algae. The spores of benthic algae also appear during the spring and summer months. Phytoplankton organisms are small and most are unicellular; but some are colonial, or are grouped together in clumps or chains. The planktonic animals span a much wider size range and include representatives from virtually every phylum found in the marine environment. Those species that spend their complete life cycle in the plankton are referred to as holoplankton, and include large numbers of copepods, as well as chaetognaths (arrow-worms), pelagic tunicates, ctenophores (comb-jellies) and siphonophores. Temporary members of the zooplankton are known collectively as meroplankton, and

Table 8.1 The broad divisions of plankton according to size.

Less than 2 μm	Ultraplankton (bacteria and viruses)
2–20 μm	Nanoplankton (mostly flagellates)
20–200 μm (0.02–0.2 mm)	Microplankton (mostly phytoplankton) also protozoans and the smallest zooplankton)
0.2–20 mm	Mesoplankton (mostly copepods)
2–20 cm	Macroplankton (large zooplankton, e.g. euphausiids, medusae, salps, arrow-worms)
More than 20 cm	Megaplankton (jellyfish)

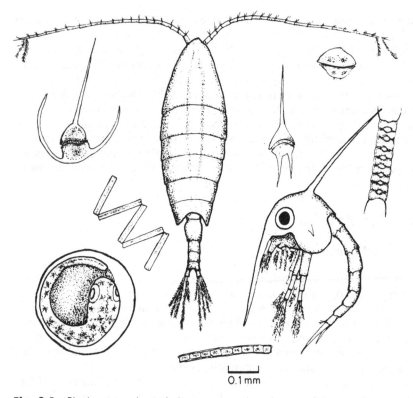

0.1 mm

Fig. 8.1 Plankton sample, including a copepod, crab zoea, fish egg, diatoms and dinoflagellates.

include fish eggs and the eggs and larvae of a wide range of benthic organisms. Some of these larvae (planktotrophic larvae) live in the plankton for several months and feed on other planktonic organisms; others (lecithotrophic larvae) remain for only a few hours or days, and rely on stored products as an energy source. Other elements of the meroplankton include bottom-dwelling mysids and ostracods that swim up into the planktonic community during the night, and adult syllid and nereid worms that aggregate in open water for breeding.

8.2.2 Staying afloat

Planktonic organisms need of course to be able to stay afloat, and often it is advantageous if they can occupy a particular position in the water column. Phytoplankton, for example, will thrive only in the photic zone, while herbivorous and carnivorous zooplankton need to position themselves close to their respective food source. The ways in which phytoplankton stay in the photic zone still have not been entirely explained, but it seems that water turbulence may be of critical importance, especially for non-motile species. Often these species, even though they have high surface area to volume ratios because of protuberances or cell shape, are still negatively buoyant and will sink in calm water. This is especially true where the cell wall or outgrowths are silicified or calcified. Cell density is also influenced by physiological processes, such as the presence of gas vacuoles, the type of reserve material present, and the ionic composition of the cell sap. Motile micro-algae such as dinoflagellates can propel themselves through the water, apparently at rates up to 48 m per day and thus, provided they are not swept away by down-currents, can keep within the photic zone. This explains the correlation that often exists between the degree of turbulence and the type of phytoplankton present. For example, at tidal mixing fronts (p. 9), the dominant organisms of the mixed side of the front tend to be diatoms, while those on the stratified side are dinoflagellates. The mixed side of the front is characterized by turbulent water, which helps to prevent the non-motile diatoms from sinking, while the stable, stratified side is occupied by dinoflagellates with their own in-built motility. Similarly, there are patterns in abundance of non-motile : motile species in relation to seasonal changes in water turbulence (section 8.3.3).

Zooplankton organisms can swim, and in addition, have various aids to increase their buoyancy. This is particularly necessary for the numerous crustaceans and other organisms with a calcareous skeleton. Almost invariably, in these cases, there are delicate elongate processes which serve to increase buoyancy by increasing the surface area to volume ratio. They probably also function as anti-predator devices. Zooplankton organisms that have no exoskeleton often have some sort of soft structure that increases buoyancy and also is involved in locomotion. The long, flattened, parapodia of the polychaete *Tomopteris*, the ciliated wing-like velum of the molluscan veliger larva and the umbrella of jellyfish are just a few examples. These adaptations reduce the rate at which zooplankton organisms sink, but do not contribute positive buoyancy. However, as with phytoplankton, the latter can be achieved in various ways. Siphonophores such as *Valella* and *Physalia* have gas-filled floats, and the pelagic gastropod *Ianthina* produces a raft of sticky bubbles to which it is attached. Medusae, salps, comb-jellies and certain molluscs have large amounts of buoyant gelatinous tissue, while fish eggs and many copepods contain significant quantities of oil or fatty material which has a similar effect.

Most zooplankton organisms, although unable to make headway against strong currents, can move quite fast considering their tiny size. Euphausiids, for example, which are relative giants amongst the planktonic crustaceans, can swim upwards at speeds of 100–400 metres per hour, while medium-sized copepods move at 30–60 m hr^{-1} and barnacle nauplii at 10–15 m hr^{-1}. Light is a strong stimulus to zooplankton, and most species, if not all, orientate themselves into their optimal level of illumination. The overall pattern is to drop downwards during the day, then to migrate up the water column at night. Some species consistently reach the surface, others position themselves at different levels in a fairly predictable order, although the actual depth will depend on the darkness of the night. It is possible that vertical migration enables individuals to select specific areas that are occupied by fewer competitors and/or more abundant food supplies, and also to move into currents flowing at different speeds, thereby exploiting new localities for food. Another theory is that their nightly ascent enables them to approach their food source under cover of darkness, when they are less likely to be seen and taken by predatory zooplankton species.

8.2.3 Successional changes in plankton communities

Distinct successional changes occur in plankton communities as a result of seasonal environmental change and pressures from grazing (Figs 8.2 and 8.3). In late winter the plankton population is at its lowest; the phytoplankton element cut back because of reduced solar radiation and low temperatures, and the zooplankton population at a low level because of lack of food and the absence of temporary reproductive stages (meroplankton) which are produced in large numbers during spring and summer. Usually the first noticeable change in early spring is the appearance of large numbers of diatoms. These are typical r-strategists (p. 69); small in size and responding to the high nutrient levels and elevated temperatures with rapid growth. In addition, being non-motile, they need turbulent water to keep them afloat. These spring blooms tend to occur in patches which, as they develop, attract zooplankton from the surrounding water. These organisms concentrate their grazing efforts on the phytoplankton patch and, while the plant cells here decline, the phytoplankton population in areas vacated by the zooplankton increases. As summer progresses there is a switch to slower-growing K-strategists with more complex nutritional requirements. Many of these are dinoflagellates, which are able to swim and so can keep afloat in the relatively stable summer conditions. In autumn there is typically an upsurge in phytoplankton abundance in response to increased nutrients released during the period of 'overturn', when stratified waters are mixed due to turbulence.

In addition to these broad changes there is also a distinct succession of species, especially amongst the phytoplankton, with one species following the

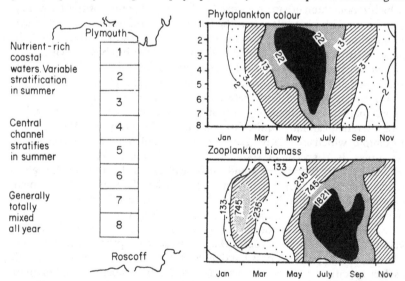

Fig. 8.2   Seasonal distribution of phytoplankton and zooplankton across the English Channel for the period 1974–1981. The phytoplankton contours show variations in phytoplankton colour (arbitrary units, giving a measure of total phytoplankton). Zooplankton contours show biomass (μg dry weight per sample). From Robinson *et al.* (1986).

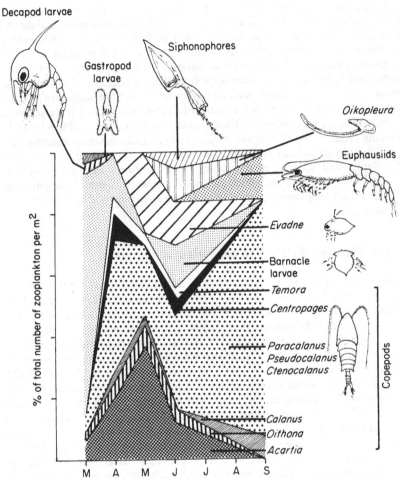

Decapod larvae

Gastropod larvae

Siphonophores

Oikopleura

Euphausiids

Evadne

Barnacle larvae

Temora

Centropages

Paracalanus
Pseudocalanus
Ctenocalanus

Copepods

Calanus
Oithona
Acartia

% of total number of zooplankton per m²

M A M J J A S

Fig. 8.3 Faunal composition of zooplankton samples taken from French coastal waters (Ushant), in the approaches to the English Channel. Hauls were taken from the bottom at 100 m depth to the surface, from March to September, 1982. Data from Moal *et al.* (1985).

other as each is successively grazed by zooplankton. There is also evidence to suggest that the phytoplankton itself, by liberating various metabolites (exocrines) so condition the water, making it more or less suitable for other species. Vitamin B12 is almost certainly implicated, and there may also be various antibiotics and other biologically-active substances involved.

Common in the spring bloom are the diatoms *Thalassiosira, Chaetoceros, Biddulphia* and *Skeletonema costatum*. The planktonic larval phases of a wide range of benthic animals enter the plankton at this time, and there is evidence to suggest that larval release in some species may be timed to synchronize with the spring phytoplankton bloom. During the summer species diversity is high with dinoflagellates such as *Ceratium* and *Peridinium* and the diatoms *Guinardia* and *Rhizosolenia* common. From August onwards there is a large meroplankton carnivore component, which represents the late developmental stages of various crustaceans. The predatory activities of these organisms cause a decline in the population of smaller, omnivorous copepods. In autumn, during the time of overturn, species composition changes again as some organisms are plunged into cooler, darker waters, while others increase in numbers in response to increased nutrient availability. Dinoflagellates are numerous, and species of the diatoms *Chaetoceros*, *Biddulphia*, *Rhizosolenia* and *Coscinodiscus* prominent.

Winter phytoplankton populations are sparse but are typified by a few dino-flagellates, and species of the diatoms *Coscinodiscus* and *Biddulphia*. In addition, benthic diatoms which normally reside in coastal sediments may be lifted into suspension as a result of winter storms and increased water turbulence.

8.2.4 Plankton distribution and indicator species

Open water appears to be a much less physically variable habitat than the sea-bed, and because of this could be expected to support a relatively uniform community. This is true to some extent, but even so, there are distinct regional and localized distributional patterns which arise in response to variations in the physico-chemical environment, and as a result of biological interactions. Gross differences occur between coastal and open sea waters, and between south-western and north-eastern areas. The classical studies by Russell (1935) revealed that particular species can be used as indicators of certain water types. Perhaps the best known are the arrow-worms *Sagitta elegans* and *S. setosa*, the former characteristic of oceanic western waters, the latter indicative of the southern North Sea and eastern English Channel. On a broader scale, a striking feature of western waters is the abundance of siphonophores, salps, doliolids and exotic Lusitanian species. At a more provincial level, in the North Sea, sixteen water masses have been identified, each with its own characteristic phytoplankton population. An analysis of zooplankton distribution in the Bristol Channel (Collins and Williams, 1982) showed four distinct assemblages of species, each of which correlated with the salinity regime and was characterized by a numerically dominant indicator species (Fig. 8.4). It is also well known from other studies that planktonic organisms are extremely sensitive to salinity, in some cases detecting differences in salt content of one part per thousand.

Just as certain holoplanktonic species are characteristic of particular bodies of water, so the meroplankton element may be similarly diagnostic. In offshore waters meroplankton is poorly represented, while in coastal areas during spring and summer there are numerous young stages of bottom-dwelling organisms, with species composition reflecting to some extent the types of benthic community in the vicinity.

Even within these recognizably distinct regional communities the plankton is seldom uniformly distributed, and samples taken only a few hundred metres

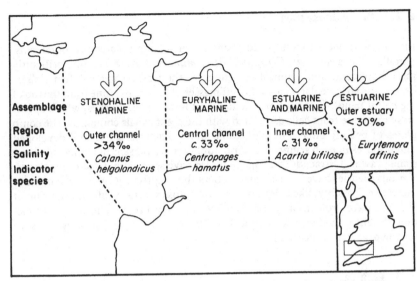

Fig. 8.4 Zooplankton assemblages in the Bristol Channel and Severn Estuary, correlated with the salinity regime. From data in Collins and Williams (1982).

apart at the same time and depth in the open sea can show distinct differences in species composition and biomass. There may be a number of reasons for this patchiness. Often it stems from localized population explosions of phyto-plankton, followed by migration of zooplankton, and then cyclical changes due to interactions between grazer and food source, or predator and prey. In other instances, for example at tidal mixing fronts, or where winds are influencing water circulation and creating vortices at the surface, it is physico-chemical factors that create the patchy effect.

Distribution patterns are complicated still further because of vertical as well as horizontal gradients and differences in factors such as light intensity, temperature, food supply and so on, which influence the phytoplankton. The phenomenon of daily migrations through the water column has already been mentioned, and there are also species that show seasonal changes in their verti-cal distribution. For example, *Calanus finmarchicus* overwinters as a develop-mental stage (the copepodite) in water as deep as 1000 m, before it matures and migrates up to surface waters in the spring. In the Celtic Sea, off south-west Britain, the water becomes thermally stratified in the summer, and this has a strong influence on the distribution of zooplankton species (Williams, 1986). Four of the most abundant species found in the colder water (about 8 °C) below the thermocline are the medusa *Aglantha digitale*, the arrow-worm *Sagitta elegans*, the copepod *Calanus finmarchicus*, and the euphausiid *Thysanoessa inermis*. The water above the thermocline reaches a temperature of 17 °C, which is at, or above, the upper tolerance limit of these species. In contrast, the copepod *Calanus helgolandicus* and the euphausiid *Nyctiphanes couchi* were found above, or diurnally migrating through, the thermocline. These are both southern species which can tolerate higher temperatures.

8.2.5 Bioluminescence

Bioluminescence is a widespread phenomenon amongst planktonic organisms. Dinoflagellates such as *Ceratium, Peridinium* and *Noctiluca* are brilliantly phosphorescent, and amongst coelenterates the warm-water jellyfish *Pelagia noctiluca* is spectacularly so, giving off masses of luminous slime from the top of the bell if handled. Many comb-jellies emit luminescent flashes, which apparently originate from cells associated with the eight canals containing the comb plates of beating cilia. The planktonic worm *Tomopteris* has well developed photophores in each segment, and photophores are present in euphausiids and some prawns. Ostracods and copepods also have special light-secreting glands. Various theories have been advanced about the adaptive significance of bioluminescence. Possibly, like colour patterns, it helps in intraspecific recognition; it also may have a role in camouflage and defence. For bacteria, however, production of light is evidently a 'chemical accident', with no particular significance for the organism concerned.

8.3 Fishes

8.3.1 Introduction

Fish are dominant amongst the nektonic organisms, and are found throughout the water column, although different species have particular preferences for certain stations. Many marine fish spend their entire lives in the upper layers of the water column, feeding on plankton or nekton. Typically, these pelagic species are strong and fast swimmers, often living in shoals, and ranging widely within their favoured feeding areas. Other fish are semi-pelagic in their habits, and are found close to the sea-bed, from which they often feed. These fish tend to have a restricted home range in comparison with the pelagic species and, although capable of fast swimming, are also adept at hovering. Because of their mobility, fish tend to be thought of as animals of open water, but there are a large number that spend most, or all, of their time on the sea-bed, and feed exclusively on benthic fauna or on animals that swim or float close by. Some of these species, including flatfish, skates and anglerfish, are quite active in short bursts, and may cover considerable ground in their search for food. Others, such as blennies, gobies and scorpionfish are much more restricted in their movements, and tend to remain in the same locality for long periods; probably many are territorial. This collection of substratum-dependent fishes are, in effect, part of the benthic community, and have been mentioned in previous chapters.

Fish differ from plankton because of their ability to make headway against waves and currents, but they still have the tendency to sink. Swimming to prevent sinking involves considerable expenditure of energy, and, like plankton, fish have evolved ways of increasing their buoyancy. Most open water fish are able to adjust their buoyancy by means of a gas-filled swim bladder, although the mackerel and its relatives lack a swim bladder and need to keep swimming in order to maintain their position in the water column. It is interesting to find that bottom-dwelling (demersal) fish generally have a swim bladder during the larval phase, but lose it prior to settling on the sea-bed.

Fish are active, mostly carnivorous animals, and their sensory apparatus and behaviour is correspondingly more complex than either planktonic, sedentary or sessile organisms. Their vision and sense of smell and taste are excellent, and in addition they have the ability to detect minute movements and vibrations in the surrounding water through their lateral line system. This enables them to navigate with the skill which is required if they are to detect obstacles, and interact with their own or other species.

8.3.2 Pelagic species

Probably, at the beginning of this century, the Herring *Clupea harengus* was the most common of all pelagic species to be found around the British Isles. Since then, as a result of massive exploitation of both adults and juveniles (whitebait), populations have declined dramatically, especially in the North Sea, where the fish is now relatively uncommon. Although larvae, juveniles and adults all live close to the surface of the water, the eggs in this species are negatively buoyant and sink to the sea-bed, where they adhere to sand, gravel or stones. Another widespread species is the Sprat, *Sprattus sprattus*, which occurs in large shoals in coastal waters, and is tolerant of low salinity waters. Adult Sprats feed mainly on copepods, while juveniles feed on diatoms and eggs and young of copepods. Sand smelts, *Antherina* spp., are also often found in waters of low salinity, again feeding on zooplankton. The Mackerel, *Scomber scombrus*, is another common member of the pelagic community, occurring for much of the year in huge shoals near the surface of the water, where it feeds on a variety of planktonic organisms.

Gobies are seldom thought of as pelagic fish, but both the Crystal Goby, *Crystallogobius linearis*, and the Transparent Goby, *Aphia minuta*, have this habit. The former is restricted mainly to offshore waters, but the Transparent Goby lives inshore and is sometimes found in huge shoals. These small, but fast growing gobies are one of the few species of fish with an expected life-span of only a year. After hatching in the autumn, they have grown to maturity by the following autumn, and after spawning is over the adults die.

There are various other pelagic species that may form shoals, but often are found singly or in small groups. The Basking Shark, *Cetorhinus maximus*, comes fairly close inshore during the summer, occasionally in shoals of up to about 50 individuals. These huge fish frequent surface waters at this time, feeding on zooplanktonic organisms such as copepods, decapod larvae and fish eggs and larvae. In winter they are known to move into deeper water, although much remains to be learnt of their behaviour and distribution. The Thresher Shark, *Alopius vulpinus*, may occasionally join the pelagic community during the summer, although rarely penetrates as far as the North Sea. The Porbeagle, *Lamna nasus*, is another of the shark family that is more common in summer than winter, and this species is known to take both pelagic and demersal fish.

There are other, even more 'exotic' fish that join the pelagic community during the summer months which are discussed in section 8.3.4.

8.3.3 Semi-pelagic species

There are numerous fish that, although living in open water, have a close affinity with the sea-bed, usually because it is from here that they obtain their food. Some tend to congregate around rocky areas, others prefer sand, or frequent a mixture of different substratum types. Probably the most familiar of these semi-pelagic species to be found in rocky areas are the wrasses. There are five species that are permanent residents of inshore waters, and which are relatively common in the south and west, but rare in the North Sea. The Goldsinny, *Ctenolabrus rupestris*, and the Ballan Wrasse, *Labrus bergylta*, penetrate further up the English Channel than the Corkwing, *Crenilabrus melops*, the Cuckoo Wrasse, *Labrus mixtus*, and the Rock Cook, *Centrolabrus exoletus*. Wrasse never move far from the sea-bed, and feed almost entirely on benthic invertebrates including worms, amphipods, isopods, crabs, and various molluscs and fish. Several species build nests of fine algae in which the eggs are deposited and guarded by the male. Ballan Wrasse and Cuckoo Wrasse also change sex from female to male, a phenomenon which is particularly noticeable in the Cuckoo Wrasse because of associated colour changes. The male has vivid blue and orange markings, especially on the head, whereas the female is brownish in colour with three dark spots on the back.

Like the wrasses, the Two-spot Goby *Gobiusculus flavescens* is a semi-pelagic species often found hovering in sheltered situations just above rocks or algae. Adults feed chiefly on pelagic copepods, while the young fish take algal-associated copepods.

A group of fish well represented in rocky areas and over broken ground are members of the cod family (Gadidae). The northern Atlantic has a greater diversity of gadoid species than any other region, and many of these are bottom-associated species; some living close inshore, others in deeper, cooler waters of the continental shelf. One of the commonest gadoids seen in shallow water is the Bib, *Trisopterus luscus*, which often forms dense shoals around reefs or wrecks in sandy areas. The Pollack, *Pollachius pollachius*, is another inshore gadoid, again generally seen in rocky areas, swimming singly or in small aggregations just above the kelp and rocks. Several other gadoids occur as juveniles and young fish inshore, but move into deeper water as they mature. For example, juvenile Cod, *Gadus morhua*, are sometimes seen in rocky areas, but adults generally form schools in much deeper water, about 30–80 m above the sea-bed.

Bass, *Dicentrarchus labrax*, often appear in the vicinity of rocks and sand during the summer, or off sand and shingle beaches, where they feed on crustaceans and fish. Grey mullets, (*Chelon labrosus* and *Liza* spp.) are common in inshore waters, and often are found in estuarine conditions. They are some of the few fish in which algae are an important part of the diet, although they also take some benthic invertebrates. The Whiting, *Merlangus merlangus*, is a particularly common fish in the North Sea, although occurs all round the coast of the British Isles. Young fish are found close inshore, and take considerable quantities of shrimp, young Shore Crabs and amphipods, as well as gobies and sand eels. However, as they grow, the fish element of their diet becomes increasingly important and includes Sprats, Bib, and even young Whiting. Thus, in effect, they change to a more pelagic existence as they mature.

Sand eels, (*Ammodytes* spp. and other species), are widespread around the British Isles, and are one of the few examples of fish that are equally at home swimming in open water, as they are burrowing into sand or shingle. In their pelagic mode they generally occur in huge shoals, while on the sea-bed, they gather into aggregations. They are an immensely important source of food for a wide range of commercial species, including Herring, Cod and Haddock, and are also taken by sea-birds such as terns and guillemots.

8.3.4 Migratory behaviour, distribution and 'exotic' visitors

The speed at which fish can swim is extremely variable, but an essential point is that they have the freedom of movement and basic ability to migrate from one area to another. For some species, vast distances are involved; for others, migration is a much smaller-scale operation. It does mean, however, that there are often noticeable seasonal changes in the distribution and abundance of fish around the British Isles. Migrations can generally be linked to a specific requirement for the species involved, particularly in relation to feeding and breeding, although the precise stimuli that trigger the migratory behaviour and steer the fish in the right direction are not entirely understood. Some of the spawning migrations of marine fish are astonishing simply on account of the distances and accuracy involved. Best known are the Common Eel, *Anguilla anguilla* and the Salmon, *Salmo salar*. The former spawns in the Sargasso Sea in the western Atlantic, then the larval forms make their way over a period of about two and a half years across the Atlantic to enter rivers where they feed and grow to adults. In contrast the Salmon spends most of its adult life at sea but migrates inshore and up river to spawn.

The migratory behaviour of Herring in the North Sea involves cross-sea movements, but is complicated because several distinct races are involved. Individuals belonging to different races occupy specific zones, and migrate at certain times to specific areas to feed or spawn.

Many other marine species show inshore/offshore migrations at specific times of year, often moving into southern and western areas of the British Isles from further south. For example young Bass, *Dicentrarchus labrax*, arrive inshore from the south-west in early spring, and the larger fish appear a few weeks later, often moving into brackish waters during the summer. In autumn most of the fish move away from inshore waters, the youngest ones being the last to leave. Mackerel, *Scomber scombrus*, make their way into the English Channel and inshore waters during spring and early summer, where they occur in huge shoals near the surface of the water, feeding on plankton. In the late autumn they descend to the sea-bed where they may feed on benthic organisms such as shrimps, amphipods, mysids and polychaete worms – although some believe that they enter a virtual state of hibernation at this time. The Black Sea Bream, *Spondyliosoma cantharus*, is a species more commonly seen in the Mediterranean, but one which, although relatively uncommon inshore in the western English Channel, makes a regular appearance during the summer months over inshore reefs at the eastern end of the English Channel, where spawning occurs. The John Dory, *Zeus faber*, is another rare visitor from the south, which may reach the North Sea, but spawns only in the south-west. It is usually found close to the sea-bed, but occasionally is seen sheltering beneath drift-wood at the surface.

A number of species rank as 'exotic' visitors to the British Isles, some being seen only once every few years, others appearing fairly regularly during the summer months. Nearly all the 'exotic' migrants come from the south, and none actually breeds in northern waters. Regular visitors include strongly swimming pelagic species such as Tunny, *Thunnus thymus*. Shoals of these fish regularly move northwards from North Africa and the Mediterranean as the temperature of the sea in northern Europe rises, and migrate into the North Sea, following shoals of fish such as Herring, Mackerel, Whiting and sand eel. Other strong swimming 'exotics' that have been recorded from around the British Isles include the Swordfish, *Xiphias gladius*, and the Wreckfish, *Polyprion americanus*. Poor swimmers such as the triggerfish, *Balistes carolinensis*, the pufferfish, *Lagocephalus lagocephalus*, and the Sun-fish, *Mola mola*, also turn up from time to time, having drifted northwards from the direction of the Mediterranean.

8.4 Other nektonic animals

8.4.1 Squids and cuttlefish

These highly intelligent animals possess many features that set them aside from the snails, bivalves and other molluscan relatives that are restricted to life on the sea-bed. Movement is achieved by precisely controlled expulsion of water through the mantle siphon, combined with the use of tentacles and lateral fins. In short bursts, cephalopods can attain speeds higher than any other marine organisms. Their vision is excellent, and they have a well developed brain which enables them to learn through experience. They also have remarkable powers of rapid colour change, which are used to good effect while stalking prey or in communication between one individual and another.

The Common Cuttlefish, *Sepia officinalis*, although well adapted for open water life, lives close to the sea-bed, and spends some of its time resting, partially buried in the sediment. When hunting, it directs small jets of water into the sand, so disturbing shrimps which it then captures with high speed precision. Other cuttlefish have similar habits, but squids such as *Loligo forbesii* and *Alloteuthis subulata* are truely pelagic, except that they descend to the sea-bed to deposit their eggs. On rare occasions, giant oceanic squid such as *Stenoteuthis* appear in inshore waters off the British Isles.

8.4.2 Sea mammals

There are only two species of seal regularly seen in British waters, the Grey Seal, *Halichoerus grypus*, and the Common Seal, *Phoca vitulina*. About 90% of the North-east Atlantic population of Grey Seals live around Britain, principally around the Farne Islands and in Scotland and Ireland. Grey Seals are seen most often during the autumn, when they come ashore to pup, with males protecting a harem of 5–10 females. They can remain submerged for as much as 20 minutes while hunting, and take fish such as sand eels, as well as crustaceans and cephalopods.

Amongst the cetaceans, those occurring in the greatest numbers are the Common Dolphin, *Delphinus delphis* and the Common Porpoise, *Phocaena phocaena*. Other dolphins and porpoises are seen periodically, and over 20 species of whale have been recorded from British waters, including the Humpback, Blue, Fin and Sei, which pass along the extreme western coastlines of the British Isles in autumn and spring. Cetaceans are extremely intelligent animals with a well developed system of communication, and complex social behaviour. They are able to produce, receive and act upon a range of acoustic signals which they also use to pinpoint prey and to navigate round objects or over greater distances. Despite this ability both to navigate and communicate, a relatively commonplace aspect of cetacean behaviour is their habit of coming inshore and stranding themselves on beaches. Strandings have been recorded for almost every species known to occur in or near British waters, yet, despite many hypotheses, the phenomenon has yet to be fully explained. Baleen whales strain planktonic crustaceans from the water, but dolphins, porpoises and toothed whales are predatory animals that feed mostly on fish and/or squid.

8.5 Interrelationships in open water

There are very few interspecific associations amongst open water species occurring around the British Isles; a situation quite different from that on the seabed, where competition for space amongst sessile organisms has led to many interrelationships between different species. The few open water associations that have been identified appear to be connected either with protection and/or food gathering. Young Whiting, *Merlangus merlangus*, associate with jellyfishes such as *Cyanea lamarcki* and *Chrysaora isosceles*, remaining close to the medusa, and within the shelter of the tentacles. Young Horse Mackerel, *Trachurus trachurus*, and Haddock, *Melanogrammus aeglefinus*, also show this type of commensal behaviour, and a number of different medusae are involved. On a smaller scale, there are various planktonic amphipods that associate with gelatinous zooplankton such as salps, medusae, siphonophores and ctenophores. The adults, especially the females, remain with their host for most of their lives, and graze on mucus or on the gelatinous tissue itself. Larvae are deposited on the host, where they proceed to feed and develop free from predation.

9

Trophic Relationships and Production

9.1 Sources of food and food webs

Energy is a basic requirement for all plants and animals, and it is not surprising to find a close link between energy availability and the type, structure and productivity of sublittoral communities. The link between the growth of algae and their energy source, light, has already been described (p. 15). The animal element of the community requires organic matter in one form or another and the distribution and abundance of species is a reflection of the type and quantity of food available, and the trophic requirements of the animals concerned. Broadly speaking there are five 'ecological' categories of food available for marine organisms: attached plants (mostly algae), attached animals, mobile animals, suspended matter (including phytoplankton, zooplankton, bacteria and organic debris), and settled detritus.

Attached algae occur almost entirely on hard substrata in shallow water, and the most important herbivorous grazers are gastropod molluscs and the sea-urchin *Echinus esculentus*. In general, living macro-algae appear to be rather under-utilized as a direct food resource, but on the other hand they are known to contribute significantly to the detrital pathway (p. 98). Attached animals such as sponges, bryozoans, hydroids and tunicates occur in enormous numbers on hard substrata, and are grazed by starfish, sea-urchins, nudibranch molluscs and other benthic animals.

Mobile animals, which occur on all types of substrata, as well as in open water, provide food for an equally wide range of other mobile species, including polychaetes, gastropods, crustaceans, cephalopods and virtually all fish. Some sessile organisms, especially coelenterates, also succeed in capturing moving prey such as shrimps, crabs and small fishes. The distribution of mobile predators relates to the availability of their preferred food, and they cluster in areas where prey species are present. Some predatory species are generalists with regard to the prey species that they take, others are specialists, or have preferences for a particular species. Often, however, they have the ability to switch from one prey species to another according to the abundance of the species concerned.

Water-borne particles of food are a tremendously rich source of food for marine organisms, whether in the form of living plankton or bacteria-coated detritus. The array and abundance of suspension feeders both on the sea-bed and in open water makes this clearly evident. Suspension feeding is of parti-

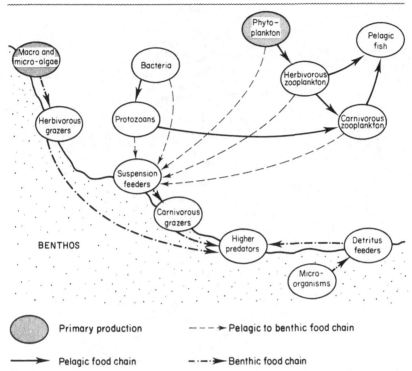

Fig. 9.1 Major components of the non-detrital food chain in coastal waters (for detrital component see Fig. 9.2).

cular importance in hard-bottom communities. For example, in a South African kelp bed, filter feeders were found to be responsible for 72% of the total animal standing stock and 77% of production (Newell *et al.*, 1982). Their food was found to consist of macrophyte particles, animal faeces and phytoplankton in roughly equal proportions. Some suspension feeders rely on wave or current-driven water movements to carry food through filtering structures or capturing devices, where it is then intercepted (passive suspension feeders). Hydroids, gorgonians, brittle-stars and feather-stars all function in this way, although the trapping mechanisms themselves differ in design. In contrast, active suspension feeders such as sponges, bryozoans, ascidians, bivalve molluscs and some of the polychaete worms generate their own currents to drive water across the filtering mechanism. Many of the pelagic animals that feed on suspended material have limb (e.g. copepods) or whole body movements (e.g. medusae and fish) that combine food gathering with locomotion. This contrasts with the situation on the sea-bed where suspension feeders are sedentary or sessile.

Benthic suspension feeders thrive in areas where there is sufficient water movement to ensure its constant renewal, and where the incoming flow carries with it an adequate supply of water-borne food. This is particularly likely in coastal areas where water tends to be nutritionally rich. There is also a correlation between the type of suspension feeder and the degree of water movement.

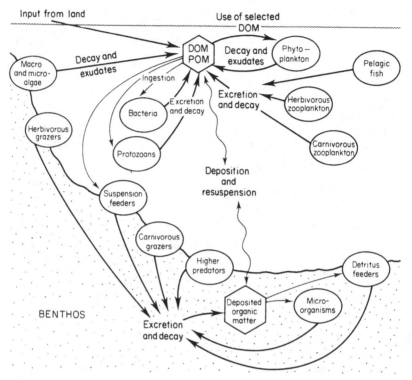

Fig. 9.2   Major components of the detrital food chain in coastal waters (for non-detrital component see Fig. 9.1). DOM = dissolved organic matter, POM = particulate organic matter.

A number of studies have shown that with decreasing water movement the population of passive suspension feeders drops, simply because food-gathering becomes very difficult for an animal in still water which is unable to circulate the water itself.

In areas where food settles out of suspension in the form of detritus there is a shift from suspension feeding to detritus feeding species which tap this energy source. Some of these worms, holothurians and crustaceans are selective feeders, picking up detrital particles, while others ingest large quantities of sediment but assimilate only the organic material.

9.2 Primary production through photosynthesis

Algae are responsible for the great bulk of primary production in the waters around the British Isles, with both macro- and micro-algae being involved. Macro-algae are limited to shallow areas where the substratum is suitably firm and light levels high. Micro-algae form films on all types of substrata, and also occur in the interstitial spaces in sediments, but a much greater range and number occur as phytoplankton. Seagrasses contribute to primary production in some areas. The relative importance of each of these components of primary production fluctuates from one area to another, but there is little doubt that, in nearshore areas, kelp forests are of major importance.

In the waters around the British Isles primary production fluctuates seasonally, and there is a marked increase in the early spring, coinciding with longer day length, increased solar radiation, higher water temperatures and a plentiful supply of nutrients. The upsurge in production is clearly evident in the form of phytoplankton blooms, the appearance of attached summer annuals, and renewed growth in perennial plants such as kelps. For example, with *Laminaria hyperborea*, new growth begins in February/April from a meristem at the base of the old blade. As the new blade is formed, so the old blade is pushed away from the stipe, and is eventually discarded, to be recycled in the form of detritus (Fig. 9.3). Over the same period, some of the annual species with short life spans pass through several generations.

Attached macro-algae continue to grow throughout the summer, and their greatest contribution to the food chain is as decaying matter, rather than living plant material. This is because there are relatively few herbivorous animals in benthic habitats. In contrast, living phytoplankton is heavily and continuously grazed by zooplankton, leaving less for the detrital food chain. However, when production is particularly high some of the algal cells die and become part of the suspended detritus, to be utilized either in the water column itself, or when they fall to the sea-bed. Phytoplankton stocks peak in spring and then fall off quite rapidly to a low summer level. However, even though phytoplankton biomass is low, productivity may still be high (see section 9.6). Indeed this seems probable, because zooplankton stocks do not decline in the same way, which indicates that they are being sustained by continued high productivity. It is known that thermal stratification prevents nutrients reaching surface waters from the benthos, so presumably recycling occurs within the photic zone.

Sufficient data now exist to assess, on a global scale, primary production by

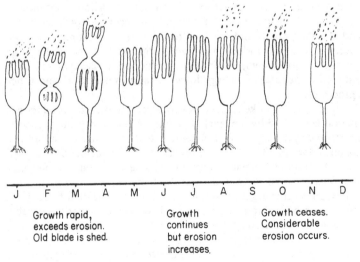

J	F	M	A	M	J	J	A	S	O	N	D

Growth rapid, exceeds erosion. Old blade is shed.

Growth continues but erosion increases.

Growth ceases. Considerable erosion occurs.

Fig. 9.3 Annual pattern of growth in the kelp, *Laminaria hyperborea*.

Table 9.1   Primary production by marine algae (from Whittle, 1977).

Province	Area 10^6 km^2	Net primary production Organic carbon	
		Mean (g m^{-2} yr^{-1})	Total (10^6 tonnes yr^{-1})
Open ocean	332.9	56.8	18 900
Upwelling zones	0.4	227.3	90
Continental shelf	26.6	163.6	4 400
Algal beds and reefs	0.6	1136.4	700
Estuaries (excluding marsh)	0.4	681.8	950
Total	361.9	70.5	25 000

phytoplankton on the one hand and attached marine plants on the other, and to compare these figures with primary productivity in other ecosystems. Production per unit area is highest in tropical seagrasses and lowest in the phytoplankton of the open ocean, but taken overall, the greatest net production is by the phytoplankton, simply because of the vast size of the open oceans (Fig. 9.4, Table 9.1).

9.3 Secondary production

All secondary producers (heterotrophs) depend for their energy on organic material built up by primary producers (autotrophs). The secondary producers comprise all the many consumer species, and often are split up according to their diet into herbivores, omnivores and carnivores. Alternatively, they may be categorized according to their place in the food chain as primary, secondary, tertiary or higher level consumers. Detrivores are also secondary producers, but use dead and decaying plants and animals, together with faecal material, as a source of organic material, rather than living organisms. Despite the huge range of secondary producers, secondary production is always less than primary production because of energy losses through respiration and other metabolic processes.

Secondary production is linked directly or indirectly to primary production, and ultimately is controlled by the level of primary productivity and the availability of the primary production. As a general rule, the planktonic community (phytoplankton and zooplankton) exports organic matter to the benthos, as well as passing it to the rest of the pelagic community. There is a transfer of energy in both directions between the nekton and the benthos. For example, some pelagic fish feed on benthic organisms, then recycle energy via their faeces both to organisms in the water column and on the sea-bed. There is also considerable interchange between the different elements within each community.

In a study of trophic relationships in shallow (10 m) well-sorted sands (McIntyre and Eleftheriou, 1968), it was suggested that the macrofauna was supported by energy input from three sources. Part came from primary production by diatoms attached to the sand grains, some from decaying macro-algae washed to the sand from neighbouring rocky areas, and the rest from primary production in the water column. Some of this production reached the benthos directly and was utilized by filter feeders, the rest took an indirect route, having passed through the pelagic food chain. It is also likely that there is an energy flow into sands through the uptake of soluble organic matter by bacteria attached to the sand grains.

In the relatively shallow waters around the British Isles secondary production in the benthos is generally high, but shows seasonal variation. In general terms, secondary production is highest over the summer months, when temperatures rise and primary productivity is at its peak. Spring phytoplankton blooms are known to trigger, after a short time lag, a corresponding increase in productivity in the deep benthic communities, (Faubel *et al.*, 1983). Some of this production is in the form of reproductive products.

Benthic communities in deeper water, where light is insufficient for primary production, depend almost entirely on an input of energy via sedimentation of organic matter. This is derived mainly from phytoplankton, zooplankton, bacteria and faecal pellets, and the supply is one of the main factors affecting production in these communities. In some areas, productivity is low because most of the primary production (phytoplankton) is utilized in surface waters by zooplankton, and even the faecal pellets and decaying remains tend to be intercepted before they can reach the sea-bed.

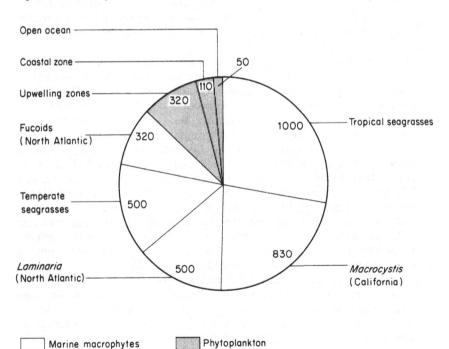

Fig. 9.4 Net annual primary production in the sea (mean figures, g per m^2 per year, from data in Dring, 1982).

9.4 Detritus and the role of bacteria

Particles of organic debris derived from living organisms are widespread throughout the marine environment and form the basis of the detrital food chain. This debris, consisting of faecal material, and dead and decaying plants, animals and micro-organisms, is a much more important source of food than might be imagined, with algae being by far the greatest contributors to the detrital food chain. This is especially so in inshore waters where most of the primary production is by macro-algae, but where there are few herbivores. Thus much of the primary production in these areas follows the detrital, rather than the herbivore, pathway of the food web (Fig. 9.2). In open and deep water, where there are no attached macrophytes, primary production is by phytoplankton, and although much of this is grazed by zooplankton, a proportion enters the planktonic detrital food chain. Precisely how much has not been determined, although it has been estimated that, under natural conditions, the grazing efficiency of zooplankton lies somewhere between 50% and 90% (Pomeroy, 1980).

A measure of the amount of detrital material reaching the sea-bed has been made by using sediment traps. In one study off the west coast of Ireland there was a sedimentation rate of 75 g m^{-2} day^{-1} during winter, falling to 20 g m^{-2} day^{-1} in summer (Shin, 1981b). The sediment collected consisted of a mixture of algal fragments, sponge spicules, diatoms, living and/or dead zooplankton, phytoplankton cells and faecal pellets, with an organic content of about 20% all year round. Many benthic animals depend wholly or partly on detrital matter for food. Others may ingest living material as well, either from suspension or from the substratum.

At the heart of the detrital pathway in the sea are countless bacteria, living in the interstitial spaces between sediment particles, on the surface of sand grains, rocks and living organisms, and suspended in the water column. Their microscopic size belies their importance, and means that they are generally overlooked by ecologists, yet it is clear that they are as critical to the functioning of the marine ecosystem as are the primary producers.

One reason why bacteria are so useful and successful is that they are able to utilize and degrade practically all types of organic matter. Much of this in the form of detrital particles, but sea water also contains dissolved organic matter (DOM) which bacteria can utilize. The part played by bacteria in the mineralization of organic detritus and release of nutrient salts has long been understood, but it is only in recent years that the importance of bacteria as producers has been recognized. In some shallow waters where production by macrophytes is high (therefore providing organic debris), and there is an additional input of detritus from the land, it has been found that bacterial production can exceed phytoplankton production. An additional feature of bacterial production is that it results in protein and mineral enrichment of organic wastes such as algal fragments, faecal material, extracellular products of phytoplankton and mucus secretions of algae and animals, which then become more attractive for animal consumption and so re-enter the food chain.

9.5 Productivity and energy flow

Productivity is a measure of the amount of organic matter (expressed as the weight of carbon) produced over a specified period of time; for comparative purposes this is usually a year. Within an ecosystem, the total productivity can be measured, or a part of it. For example, primary productivity or the productivity of a particular community or species. The biomass or standing stock of a species does not give a measure of productivity but simply the weight of that species present at a particular moment in time, or the mean weight over a period. It is often assumed that a large biomass signifies high productivity, but in fact this is not necessarily so, and quite the reverse can be true. Productivity is a measure of the rate at which organic material is produced, and some organisms with a large biomass add new material very slowly, generally because the energy input is largely used up in respiration. Thus their productivity to biomass ratio is low. Other, smaller organisms (e.g. bacteria, phytoplankton) may be much more efficient, and the productivity to biomass ratio is then high. On average, annual production by phytoplankton is somewhere between 15 and 45 times their standing biomass. In contrast, that of larger, longer-lived organisms such as fish, is usually less than their standing biomass. A study of benthic production of three bivalve communities in the Bristol Channel revealed that annual production in the *Abra* and *Modiolus* communities just exceeded the mean annual biomass, while in the *Venus* community, productivity was lower (Table 9.2). Many other organisms in addition to those which gave the communities their names were involved in production. For example, the *Modiolus* community was in fact dominated by the brittle-star *Ophiothrix fragilis*, and it was thought that the high productivity was due to the greater proportion of suspension feeders in this community than in the others.

In recent years attention has been turned to broad-based studies which examine the functioning of entire ecosystems. This type of approach has important practical applications, especially in relation to fisheries management, where details of productivity and energy flow need to be known. An ecosystem is often examined step by step through the food web, and an average figure for efficiency quoted as the move is made from one trophic level to the next is about 10%. However, a practical problem with this type of approach is that many organisms do not fit neatly into one particular level. It is thus more relevant to study the relationship between ingestion and growth (gross growth efficency) of

Table 9.2 Annual production (P), mean annual biomass (B) and production/biomass ratio (P/B) of benthic communities in the Bristol Channel. Adapted from Warwick (1984), and George and Warwick (1985).

Community	Production (P) (g dry wt m^{-2} yr^{-1})	Biomass (B) (g dry wt m^{-2})	P/B
Venus	25.8	45.8	0.56
Abra	14.2	11.1	1.3
Modiolus	34.1	24.5	1.4

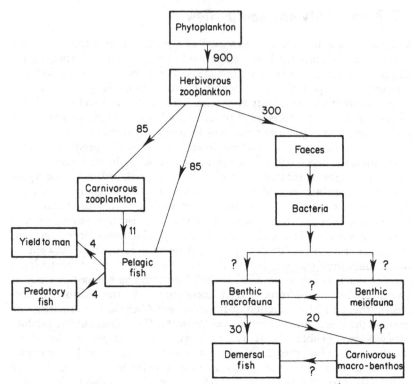

Fig. 9.5   Food web and energy flow (kcal per m² per year) in the North Sea. From the model put forward tentatively by Steele (1975).

the organisms concerned. Macro-invertebrates, for example, have been found to have gross growth efficiencies of about 20–30%, and laboratory experiments have shown that bacteria can have efficiencies in excess of 50%.

A model for the North Sea has been tentatively constructed by Steel (1975), using field data collected from many sources as a basis for his estimates (Fig. 9.5). The point is made, however, that there are several uncertainties or unknown aspects to this model, for example, regarding the relationship between the benthic meiofauna and other elements of the benthic biota. Often models such as this produce almost as many questions as answers concerning the dynamic functioning of ecosystems, but as more data are pooled, more accurate, but complex pictures emerge. Considerable effort has to be put into sampling the various elements of the ecosystem, but it is possible, as Brylinsky (1972) has done for the English Channel (Fig. 9.6), to make use of existing data on standing crops, and to calculate energy flow from more recent information on calorific values and efficiencies.

The main difficulty is ensuring that all pathways are taken into consideration and accurately assessed. For example, it is likely that zooplankton production in coastal waters is underestimated because temporary larval plankton released by benthic animals often remains close to the sea-bed, and so is inadequately

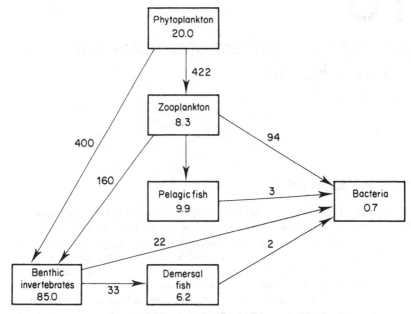

Fig. 9.6 An estimate of production in the English Channel. Standing crops (figures in boxes) are in kcal per m², energy flow in kcal per m² per year. From Brylinsky (1972).

sampled. One study has shown that of the total larval polychaete population in the water column above a rocky sea-bed, about 80% were captured in the bottom sample, the rest in mid-water or surface samples (Wilson, 1982). Similarly, macrophyte production which contributes decaying matter to the system, and transfers energy from the coastal fringe to the body of the sea has to be taken into account.

10
Use and Management of Coastal Seas and their Resources

10.1 Introduction

The seas around the British Isles have been put to man's use for many years, but the twentieth century has seen an upsurge in exploitation and pollution, coupled with a diversification in other activities that impinge on marine ecosystems. Fortunately, the resilience of the marine environment means that much has gone on without significant damage being caused. However, this does not mean that the sea has an infinite capacity to recover from all that is dumped into it or taken from it. Even now, it is known that some activities are detrimental and that certain species and habitats are more vulnerable than others to pollution, disturbance and exploitation.

Until fairly recently activities in or at the fringe of the sea were allowed to continue with little control or consideration, but there are now growing calls for a coherent policy on coastal zone and resource management, as well as planned strategies for waste disposal. This applies to many countries, not just the British Isles.

The most recent appraisal of marine-related activities and their impact was put together as part of the United Kingdom's response to the World Conservation Strategy (Shaw, 1983). This report pinpoints a number of activities which cause serious and often irreversible damage to living resource conservation in the marine and coastal zone, and goes on to discuss how the damage might be prevented. One problem with this type of approach is that it overlooks cumulative effects where a particular area, habitat or species can withstand a single isolated activity but is damaged in some way by a combination of several different ones. This chapter provides a broad picture of the uses to which our seas and marine resources are being put, the associated problems, and the types of conservation, management and anti-pollution measures either planned or in operation.

10.2 Commercial fisheries

Fish, crustaceans and molluscs caught from the wild form the basis of our fisheries, but mariculture is becoming increasingly important. Three rather separate issues have to be considered when assessing the impact of fisheries. One is the question of over-fishing; the others are the environmental side-effects of fishing gear on the one hand and mariculture activities on the other.

10.2.1 Over-exploitation

Over-exploitation of fish stocks around the British Isles has been well documented for the past 40 years or more, with stocks in the North Sea showing the most dramatic declines. The status of fish stocks around the British Isles, assessed by Cushing (1980), is shown in Fig. 10.1. Migratory species such as the Salmon, that are taken in fixed nets as they move up estuaries, are also heavily fished. Sharks, rays and other elasmobranchs may be vulnerable to over-exploitation because they are slow breeders with a low fecundity. The Common Skate, *Raja batis*, which does not mature until it is about 11 years of age and produces few eggs, has been virtually eliminated from the Irish Sea as a result of over-fishing. Similar dangers face the Basking Shark, another slow-breeding species which is exploited for its liver oil, fins and other by-products. Little is known about the migratory patterns, reproductive biology and population dynamics of these huge fish, yet the fishery continues without any controls. Often it is not only the principal target species that are affected. Many juvenile and 'trash' fish of little or no commercial value for food are caught, either to be discarded, or to be used for industrial purposes such as the manufacture of fish meal.

Molluscs and crustaceans have also come under increasing pressure in recent years, and a greater variety of species is now being taken, often to meet the demands of overseas markets. Queen scallops, shrimps, prawns and Norway Lobster, *Nephrops norvegicus*, are taken by trawling; scallops and oysters by

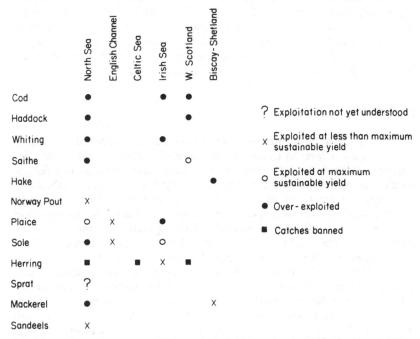

Fig. 10.1 State of commercial fish stocks in British waters in 1980 (Cushing, 1980).

dredging, and crabs and lobsters by potting. Crawfish and lobsters are also taken by tangle-netting and scuba diving. There is little doubt that over-exploitation is occurring in several areas, especially for crawfish in south-west England, but the status of populations of commercial species is not adequately known.

The sea-urchin *Echinus esculentus* is one of the relatively new commercial species and is taken by scuba divers both for food and for the curio trade. Again, there are dangers of local depletions in areas where the collectors operate.

10.2.2 Damaging side-effects

Fishing has a direct impact on populations of exploited species, but the gear itself can have damaging side-effects on other organisms, and on benthic habitats. Heavy trawl nets disturb the sea-bed as they are dragged over it, but probably the most damaging are scallop dredges, which have heavy spikes that plough through the sediment, destroying and disturbing large numbers of benthic organisms.

The practices associated with mariculture, an increasingly popular industry involving a growing range of species, may also have repercussions which are unfavourable to established communities. Alien species introduced for culturing purposes may become established in the wild, as has happened with the clam *Mercenaria mercenaria*, although in this case there do not appear to be detrimental ecological side effects. Bivalve farming often involves other species being excluded, and this may upset the ecological balance of the area. Where fish are kept in enclosures and fed intensively with artificial foods, organic matter tends to build up, and eutrophication occurs. Often these cages are in the semi-enclosed waters of sea lochs, and communities previously thriving in the sediments below and around the cages may be devastated as a result of the anoxic conditions that ultimately follow eutrophication.

10.2.3 Management

Management policies for marine fisheries are required in order to keep exploitation at such a level as to obtain maximum yields of commercial species, without causing declines in stocks. It is also necessary to ensure that fishing practices cause minimal incidental damage to other species and to habitats.

The need for international co-operation to avoid over-exploitation is particularly critical for migratory species, and in areas such as the North Sea where all the bordering nations want to exploit fish populations to the full. In the north-east Atlantic the International Council for the Exploration of the Seas (ICES) assesses fish stocks and passes the information on to the north-east Atlantic Fisheries Commission, which allocates quotas to each of the fishing nations. The fishery authorities in each country then regulate the industry by putting limits on the number of fishing boats that can operate, and the amounts and size of fish that can be landed; and by designating closed seasons and restricted areas. The fishery for crabs and lobsters is controlled mainly through regulation of minimum landing size. There are few controls on capture of other crustaceans and molluscs. In theory these restrictions should prevent over-exploitation, but there are many practical problems involved, particularly with regard to enforcement. In recent years, declines in catches of lobster have led to stock enhancement trials, involving the placement of thousands of tagged juveniles in selected areas of the sea-bed.

10.3 Waste disposal and pollution

Many waste materials are deliberately deposited in the sea; either being dumped, or directed through outfalls. Some filter in by accident, for example from rivers and from the atmosphere. The sea has always been seen as a convenient place for disposal of wastes, and it is often assumed that the materials will be degraded, diluted and dispersed so as to be neither a nuisance nor an environmental threat.

10.3.1 Sewage and other organic wastes

Most of the organic wastes disposed of in coastal waters are in the form of sewage and are from domestic sources, but industries such as brewing, wood-pulping, food processing and agriculture also produce organic wastes that find their way into the sea. Much of the sewage is discharged into estuaries and coastal waters with little or no treatment, and only about a quarter of all pipes at present have an outlet which opens more than 100 m beyond the low tide mark. About 10 million wet tonnes of sewage sludge is disposed of annually in British coastal waters (Lack and Johnson, 1985), 60% of it in the North Sea.

These biodegradable wastes provide energy in the form of organic matter and nutrients, and in moderation, increase productivity in the sea, just as fertilizers do on land. In these cases there is enrichment of the local flora and fauna which is evident right up the food chain. Estuaries in general tend to be productive because of the organic input, and the bird-feeding grounds associated with sewage outfalls are well known. There may also be a beneficial effect on fisheries, but all these increases in production are seen only where dilution and dispersion are adequate to prevent over-loading. If the input exceeds the capacity of receiving waters then problems arise. Eutrophication occurs, and the huge production of phytoplankton and macro-algae excludes most animal life and causes a build-up of decaying plant material. This is broken down by bacteria, with oxygen being used in the process, and in over-loaded areas excessive bacterial activity leads to the development of anoxic conditions, so inhibiting aerobic organisms and encouraging the growth of anaerobic bacteria. When this stage is reached the biota is put under considerable stress and only certain well-adapted species survive. Areas immediately adjacent to outfalls and estuaries with a low flushing rate are particularly likely to be faced with both these problems and, as a result, to have an impoverished fauna and flora. In addition to the problems of oxygen starvation, benthic organisms are in danger of being smothered by the sheer volume of particulate matter as it settles out on to the sea-bed. Suspension feeders are particularly vulnerable because of their delicate filtering structures.

The response to organic pollution always tends to follow a characteristic pattern as the biota is faced with increasing stress. Initially species diversity declines although biomass remains high. For example, the rocky sea-bed around sewage outfalls is often dominated by enormous numbers of the mussel, *Mytilus edulis*, and by the predatory starfish, *Asterias rubens*. In soft substrata affected by organic enrichment the polychaete worm *Capitella* is similarly extremely abundant. As organic pollution increases species diversity falls still further, and biomass also declines. In extreme situations nothing but bacteria remain.

There is growing evidence that the incidence of viral and bacterial diseases in fish and shellfish is much higher in areas receiving sewage sludge contaminated with pathogenic organisms. These diseases are not generally passed on to man, but greatly reduce the market value of the affected stock. Sewage can, however, be a direct health hazard to man, and considerably reduces the recreational potential of coastal areas. It is often claimed that around 90% of the bacteria contained in sewage will die within 4 hours of release into the sea, but in reality the time varies enormously and often is closer to 24 hours. The biggest bacterial killer in the sea is light, especially ultraviolet light, and so the survival of bacteria relates very much to this factor. In murky waters and/or on overcast days bacterial levels may become dangerously high because the bacteria are killed at a slower rate than in clear water on a bright sunny day.

Sewage rarely consists entirely of organic matter, but is usually contaminated with heavy metals, detergents, petroleum residues and a variety of other chemical substances.

10.3.2 Inert, inorganic wastes

Inert inorganic wastes deposited in the sea, even though non-toxic, have damaging effects because they change the nature of the sea-bed, smother benthic organisms and increase turbidity, thereby reducing light penetration and cutting primary productivity. Collieries and power stations produce large quantities of such wastes. For example about one million tonnes of pulverized fly-ash from the Blyth Power Station in Northumberland is dumped off the adjacent coast each year (Bamber, 1984). This fine particulate matter, which consists predominately of silicon dioxide with oxides of iron and aluminium, has caused an increase in the silt fraction of the sediments, and consequent changes in the benthic communities. Numbers of macrofaunal individuals and species, and overall biomass are depressed throughout the spoil ground, over an area of 43 km^2, with the greatest depression at the ash-dumping centre (Fig. 10.2).

The china clay industry in Cornwall produces about 1.5 million tonnes of kaolin and mica waste annually, about 75% of which is deposited in St Austell Bay and Mevagissy Bay. Here it is estimated to cover about 60% of what was originally a coarse-sediment sea-bed. The bivalves *Venus fasciata* and *Dosinia*

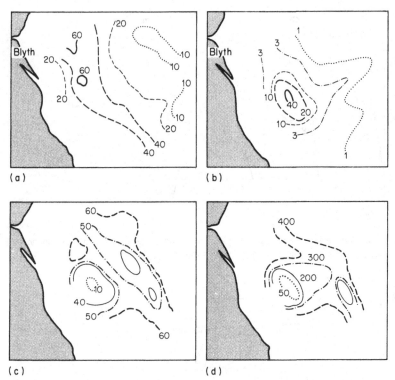

Fig. 10.2 Some physical and biological characteristics of a pulverized fuel ash dump site off the Northumberland coast (from Bamber, 1984). (**a**) Contours of silt % by weight in the sediments. (**b**) Contours of pulverized fuel ash content ($N \times 10^6$ particles per gram). (**c**) Contours of numbers of macrofaunal species (measure of faunal diversity). (**d**) Contours of macrofaunal individuals per $0.2\ m^2$ (indication of biomass).

exoleta have been eliminated from these areas, and the silty deposits now support a completely different fauna.

A high proportion of benthic organisms have delicate respiratory and feeding structures that are easily clogged by inorganic (and organic) material suspended in the water column. Increased loads of suspended particulate matter have been shown to cause reduced growth rates and mortalities in a range of animals, including bivalve molluscs, the prosobranch *Crepidula fornicata* and ascidians. Ascidians are unable to select particles or produce pseudofaeces and appear to be particularly sensitive (Robbins, 1985). In experiments with *Ciona intestinalis* and *Ascidiella scabra* moderately elevated levels of suspended material (25–170 mg 1^{-1} of Fuller's earth (a commercially available diatomaceous earth)) caused significant decreases in growth, and highly elevated levels (over 602 mg 1^{-1}) caused 100% mortality within about 3 weeks (Figs 10.3 and 10.4).

Enormous quantities of dredging spoil from harbours and shipping channels are dumped at sea. In England and Wales alone 28 million tonnes are dumped at 60 designated offshore sites (Clark, 1986). This spoil has similar effects to the particulate wastes described above, but in addition is often anoxic

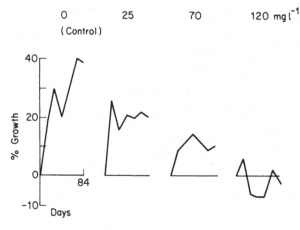

Fig. 10.3   Percent growth (volume) of *Ascidiella scabra* exposed to various concentrations (mg l⁻¹) of Fuller's earth. From Robbins (1985).

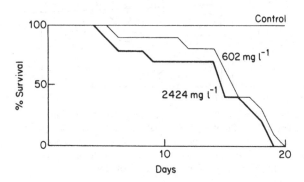

Fig. 10.4   Survival of *Ciona intestinalis* exposed to high concentrations of Fuller's earth. From Robbins (1985).

and also contaminated by heavy metals, so introducing additional problems (see section 10.3.5).

Another source of non-toxic waste is the vast assortment of refuse dumped overboard from ships, or washed into the sea from land. This includes tin cans, glass containers, nets and ropes, tyres and plastic items of all shapes and sizes. This type of rubbish may cause injuries to fish, marine mammals and birds, but the main problem is that it is an unnecessary eyesore.

10.3.3 Oil and other petroleum products

Significant quantities of finely dispersed oil and petrol reach the sea from urban and other terrestrial sources, but the major contributor of liquid oil is the petroleum industry. In the past oil tankers used to discharge their oily ballast and tank cleaning water directly into the sea, where it readily formed slicks. New regulations requiring dilution of the discharge were introduced, followed by procedures to retain the oily fraction of the tank washings, but the discharges still have not been entirely eliminated. There is also danger from accidental spillages from tankers, well-heads, and pipelines. Oil on the surface of the sea kills sea-birds by choking them and coating their feathers, and also causes heavy mortalities of fish eggs and larvae. When a large quantity of oil is washed onto the shore it can devastate intertidal life, but much depends on the susceptibility of the organisms present. At one time oil spills were invariably treated with dispersants, but these in themselves proved to be damaging, and were often relatively ineffective. When left alone the major toxic fractions in the oil evaporate within a few days, and the remaining fractions dissolve or are emulsified, and are gradually degraded by micro-organisms. The policy now is to try and contain the oil spill in order to prevent it from reaching the coast or areas where there are major populations of sea-birds.

Another type of oil-related pollution occurs around oil production platforms. There has in recent years been a general trend of increasing use of oil-based drilling muds which produce cuttings contaminated with diesel and other oils. In 1983 about 150 of the 223 wells in operation in the North Sea were drilling with oil-based muds, and discharging an estimated 7700 tonnes of diesel oil and 10 400 tonnes of alternative base oils (Davies *et al.*, 1984). Discharge of cuttings is known to cause physical smothering of the biota, especially within a radius of about 500 m from the rig. Oil concentrations between 1000 and 10 000 times background level (over 10 000 ppm total oil) have been found within 250 m of platforms using oil-based drilling muds, with the concentrations returning to background levels within about 3000 m. A combination of the smothering effect and the presence of hydrocarbons, leads typically to a severely depleted central zone, and a gradient of increasing species diversity moving outwards from this zone. Finally, by about 1000–3000 m (depending on many variables) the benthos has returned to normal (Figures 10.5 and 10.6).

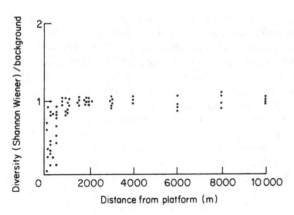

Fig. 10.5 Variation in species diversity (relative to background) with distance from the production platform (combination of results from 5 North Sea oilfields). From Davies *et al.* (1984).

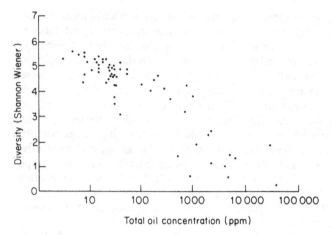

Fig. 10.6   Relationship between species diversity and total oil concentration in sediment around production platforms (combination of results from 5 North Sea oilfields). From Davies *et al.* (1984).

10.3.4 Short-lived toxic wastes and heated effluents

A range of toxic chemicals, including acids, alkalis and phenols are pumped into the sea but are rapidly rendered harmless by being diluted or neutralized. Often these substances are discharged as a heated effluent, and the combination of heat and toxicity has a devastating localized effect adjacent to the discharge pipe, with few, if any, species surviving.

10.3.5 Persistent toxic wastes

Coastal waters receive a significant input of non-degradable wastes which have toxic effects at comparatively low concentrations. They include the heavy metals cadmium, mercury, arsenic, copper, chromium, lead, nickel, and zinc, as well as organic compounds such as lindane, DDT, polychlorinated biphenyls (PCBs) and 'drins' (the pesticides aldrin, dieldrin and endrin). A review by Clark (1986) of the toxicity of these substances illustrates the widely differing effects according to the precise form of the pollutant, the type of exposure and the organism concerned. These substances are included in the European Community (EC) Dangerous Substances Directive, and loads entering coastal waters have been monitored in recent years (Table 10.1). The highest input is from river discharges, but sewage and industrial outlets also contribute to the load (O'Donnell and Mance, 1984). One of the problems is that dispersion of these toxic wastes is often uneven, and estuaries tend to act as 'reservoirs' often with the metals or other substances bound to particles in the sediment.

In Liverpool Bay, where several million tonnes of sewage sludge are dumped each year, the concentration of metals in the sediments has been rising. For

Table 10.1 Total mass of certain wastes discharged to UK coastal waters from rivers, and sewage and trade outfalls. Adapted from O'Donnell and Mance (1984).

	Flow ($\times 10^3$ Ml per year)	
Sewage	3454	
Trade wastes	1688	
	tonnes per year	
Cadmium	73–101	} List I EC Dangerous
Mercury	20–33	} Substances Directive
Copper	1423–1678	
Lead	985–1495	
Zinc	6497–6642	List II EC Dangerous
Chromium	1568–1971	Substances Directive
Nickel	963–1179	
Arsenic	309–354	
	kg per year	
'Drins	66–358	
DDT	133–710	List I EC Dangerous
Lindane	983–1101	Substances Directive
PCBs	83–662	

example, peak concentrations of lead rose from 220 parts per million (ppm) in 1976 to 1000 ppm in 1980, indicating that natural dispersion is insufficient to prevent accumulation on the sea-bed. In the Thames estuary, the heavy metal and organic content of the 5 million tonnes of sewage sludge dumped each year is having localized effects on marine life. Concentrations of heavy metals and organic substances are 5–8 times higher than background level, and have placed the natural fauna under stress, promoting the growth of pollution indicator species that are able either to utilize or tolerate the elevated pollutant levels. The extent to which organisms can regulate levels of heavy metals varies. Bryan (1976) suggests that essential metals such as zinc and copper are fairly well regulated in decapod crustaceans and fish, but not in other marine organisms (Table 10.2) while non-essential metals such as cadmium and mercury are less well regulated. In Restronguet Creek in the Fal estuary, levels of copper in the sediment are about 3000 ppm, as a result of contamination with mining wastes for about 200 years. Concentrations of copper in a variety of marine organisms from the creek are shown in Table 10.2. This type of exposure may lead to a certain degree of tolerance in specimens from highly contaminated situations, but prevents colonization by sensitive organisms, such as larval bivalves.

An overall problem with these wastes is that, in addition to directly affecting growth and physiological processes in marine organisms, they accumulate in the environment, and in the organisms themselves (bioaccumulation). In this way they enter food chains and become concentrated in animals tissues with each step up the chain (biomagnification). Filter-feeding bivalves are a common entry point, especially since these animals often dominate estuarine habitats

Table 10.2   Metal concentrations in species from a chronically contaminated site (Restronget Creek, south-west England) and more normal areas. Adapted from Bryan (1976).

		Copper concentrations $(mg\ g^{-1}$ (ppm) dry weight)	
		Contaminated site	Normal site
Platichthys flesus (flounder)	liver	118	60
	viscera	28	12
	remainder	4	3
Carcinus maenas (crab)	whole	191	77
	hepatopancreas	2540	163
	gills	341	130
Corophium volutator (amphipod)		499	96
Nereis diversicolor (ragworm)		1140	22
Nephthys hombergi (polychaete)		2120	18
Scrobicularia plana (bivalve)		111	13
Ostrea edulis (oyster)		3870	488
Fucus vesiculosus (brown alga)		2780	5

where levels of pollutants are particularly high. Higher predators, including fish, sea-birds and man are then exposed to food with large concentrations of these toxic substances. Partly for this reason, the effects of pollution may not be noticed for some time because delayed and/or sub-lethal effects such as these may not be readily detectable. In addition, unless pollution is chronic, the response time of communities is often slow and environmental damage difficult to detect above the natural background variability. The use of indicator species that are particularly sensitive to pollutants is a useful monitoring tool in this respect because they act as 'early warning' devices. The mussel, *Mytilus edulis*, is often used, and a number of physiological and other processes can be measured in order to ascertain the degree of stress to which the animals are being subjected. This type of approach can give a more direct and immediate indication of pollution effects, as illustrated in a study of the biological effects of sewage sludge at a licensed site off Plymouth. A number of general and specific indices of well-being in mussels were evaluated, and a range of responses found that correlated with site and depth. For example, according to the scope for growth (SFG – a measure of a group of physiological responses including energy assimilation, respiration and excretion) all the mussels except at site 1 were stressed (Fig. 10.7b). SFG was lowest in mussels from the edge of the dumping site and highest at site 1, indicating a reduction in stress further from the dumping site. Similarly, assessment of stress at a cellular level, by measurement of lysosomal stability, revealed that all the animals in the upper cages (4 m below the surface) were stressed (low stability), and that the lowest stability (greatest stress) was at the centre of the dump site. Of the animals from the lower cages (4 m above the sea-bed), only those at the centre of the dump site were stressed (Fig. 10.7c).

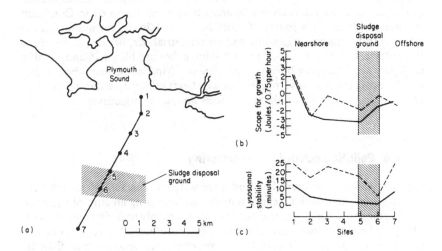

Fig. 10.7 (**a**) Position of sludge disposal ground off Plymouth; (**b**) scope for growth and (**c**) lysosomal stability of *Mytilus edulis* exposed at sea in upper cages (———) and lower cages (-----). From Lack and Johnson (1985).

10.3.6 Incineration at sea

One way of disposing of hazardous wastes it to incinerate them at sea. Around the British Isles most of this takes place in the North Sea, and wastes include heavy metals, and organochlorine compounds such as pesticides and PCBs. Incineration does not solve the problem of safe disposal of these hazardous compounds because toxic solids are still produced and heavy metals are not destroyed but merely vaporized. Around 66 tonnes of arsenic entered the sea in this way from incinerator ships in 1981 (Caufield, 1984). Incineration at sea is carried out under the rules of the Oslo Convention, which includes a commitment to phase out and ban the practice altogether.

10.3.7 Nuclear wastes

Most of the radioactive materials entering coastal waters are low or intermediate level wastes from nuclear power stations and reprocessing plants. The main causes for concern are Caesium[137], which does not occur naturally and remains in solution, together with Plutonium[239] and Ruthenium[106], which adsorb to fine particles and accumulate in sediments on the sea-bed. All these materials have a long half-life, and so remain as contaminants for many years. There has been considerable public disquiet about these discharges, especially since it became known that the Irish Sea is apparently the most radioactive in the world. More than 95% of the radioactive wastes discharged into the Irish

Sea since 1957, when Windscale (Sellafield) became operational are still in the sediments within 50 km of the discharge pipeline. This is because they have become bonded to sedimentary materials, and so have accumulated on the sea-bed. Some of these radioactive sediments have been deposited in Cumbrian estuaries, and will be a source of radiochemical pollution for many years to come, both for marine organisms and for terrestrial life, because some of the fine sediments from the mudflats and saltmarshes are blown inland, joining radioactive aerosal spray blown in from the sea. Most of the fears about radio-activity are extrapolations from the known effects of radiation damage to man; very little is in fact known about its effects on marine organisms.

10.3.8 Pollution control and monitoring

Disposal of wastes in coastal and offshore waters is controlled through both domestic and international legislation, and an increasing amount of pollution monitoring is now being carried out in order to comply with these regulations. Many of the new control measures have been introduced as a result of consistent pressure on governments from environmental, amenity and fisheries interests. However, there are still fears that standards of pollution control and monitoring are too low, and that many wastes continue to be disposed of at sea without adequate safeguards.

Control of pollution in estuaries and coastal waters was introduced in 1984, under Part II of the Control of Pollution Act 1974. Under this legislation, applications to discharge effluent have to be advertised publicly, and Water Authorities are obliged to hold a register for inspection by the public. However, improvement is not guaranteed because exemptions and 'deemed consents' allow various polluters to side-step the legislation. With regard to health hazards created by sewage, the British Government is also meant to ensure that EC standards for 'Eurobeaches' are met. This issue has been avoided to a great extent by considering only very few beaches as 'bathing beaches', but after substantial public pressure, the Government now insists that Water Authorities test the bacteriological quality of many more areas. Licences to dump sewage at sea are granted annually, in accordance with the Dumping at Sea Act, 1974.

The British Government also has to comply with a number of EC directives and international regulations which should lead to better pollution control. For example, when levels of mercury in fish from Liverpool Bay rose to between 0.25 and 0.28 parts per million, dangerously close to the limit of 0.3 ppm laid down by EC regulations, the Government renewed the licence of the Water Authority for sea disposal of sewage sludge in 1986 only on the condition that the trace amount of mercury was reduced by 30%. Many species of marine fish have only a limited ability to excrete mercury and so concentrations in the tissues increase as the fish grows (Rivers *et al.*, 1972).

Assessment of the fate and impact of wastes reaching the marine environment involves studies of dispersion, adsorption and deposition on the sea-bed or on suspended particles, re-release or resuspension, and biological uptake. The presence of wastes in the environment or biota does not *per se* mean that damage is being caused, and part of any impact assessment also requires both short and long-term studies of the biota. The determination of lethal and sub-lethal effects of every pollutant (in every form and combination) on all possible target organisms through toxicological tests is clearly an impossible task, and a combination of such studies with monitoring in the field is generally used as a basis for setting allowable limits. Monitoring is particularly valuable because it can detect the cumulative impact of a single pollutant, as well as synergistic effects of several pollutants acting together. Disturbance and stress caused by pollution can be investigated by examining the abundance and type of species occurring at a particular site. A major difficulty is that any analysis of this type also has to take into consideration naturally-occurring environmental stresses, and it is not always an easy task to separate one from the other. Useful information can also be collected through the use of biological indicator species known to be sensitive to certain pollutants.

10.4 Development in the coastal zone

The coastal zone is altered and developed for many purposes, including shipping, industry, housing and recreation. Much of the impact is directed onto coastal lands or the intertidal zone, but inevitably there are repercussions in the sublittoral. Harbour construction and reclamation are the most extreme because they lead directly to habitat loss, and estuaries have been particularly badly hit in this respect. This is especially serious considering the importance of estuaries as highly productive areas supporting a rich array of both commercial and non-commercial species. There are also many other activities, including construction of piers and coastal defences, cable laying and dredging of shipping channels which are likely to cause habitat change or degradation. The main reasons why change in benthic communities occurs is because of alterations in the patterns of water movement and sediment deposition, and through increased turbidity. On the other hand there may be some beneficial effects. A dive below Brighton Pier, for example, illustrates clearly how the pier piles, by providing a hard substratum in an area dominated by shingle, have led to diversification of both the benthic and pelagic biota. Similarly, it has been suggested that there could be considerable benefits to commercial fisheries if disused oil production platforms were not removed, but were turned instead into artificial reefs (Ralph, 1986).

Around 20 million cubic metres of sand and gravel are extracted every year from European waters, with much of the material coming from the North Sea (Lee and Ramster, 1981). This is an extremely destructive process, especially when mixed sediments are dredged, because it causes profound habitat changes as a result of disturbance during dredging, removal of the coarser grades of sediment, and deposition of fine material. Recovery of the fauna in dredged areas is slow, and there may also be indirect damaging effects on commercial fisheries, for example through destruction of Herring spawning grounds.

Many different government departments and authorities have power to control activities below the low tide mark, and a number of these, for example Water Authorities, Port Authorities and the Ministry of Defence, can also carry out developments without needing any form of approval. Environmental impact studies are not mandatory at present for marine-related development projects, and the area beyond the low tide mark is free from planning permission. Even where projects do include an environmental impact assessment, there is no guarantee that recommendations will be acted upon, and there is no single Ministry within the Government with special responsibility for marine matters. Another practical problem is that gaps in our knowledge of the functioning of elements of the marine ecosystem makes it difficult to make accurate predictions of the likely outcome of some of the activities. However, permanent monitoring sites are now being established in several sublittoral areas in order to study natural and man-induced changes in the physical environment and in the functioning and structure of marine communities.

10.5 Recreation

The sublittoral is much less prone to disturbance from direct recreational activities than the seashore itself simply because it is less accessible, and is removed from the effects of bait-digging, overturning of rocks, and trampling by visitors. However, scuba divers can have an impact through selective collecting both of edible species and of curios such as the sea-urchin *Echinus esculentus* and the sea-fan *Eunicella verrucosa*. Even though this is not carried out at a commercial level, the cumulative effect of collecting by thousands of divers can be a problem, especially in popular areas. The environmental impact of sea-angling is largely unknown, but catches as a whole are obviously miniscule in comparison with commercial fisheries. However, certain species, for example Bass and sharks, are caught in relatively large numbers, and this may affect populations. One clearly damaging outcome of sea-angling is that discarded hooks and lines are a hazard to sea-birds.

10.6 Protected areas

In 1981 it became possible for the first time to establish statutory marine nature reserves; until then protection could be given only to areas down to the low water mark. These reserves will be administered by the Nature Conservancy Council, and are intended to protect representative sublittoral areas with especially interesting marine flora and fauna. They will also be important for education and research. Byelaws will govern some of the activities that can be carried out within the reserves, but certain statutory bodies and 'authorities' are exempt. It was 5 years before a reserve was designated, mostly because of opposition from powerful lobbies such as the fisheries authorities who were anxious not to have their activities curtailed.

Several voluntary conservation areas have been successfully established through the initiative and commitment of local people. Every effort is made to retain the natural interest and beauty within these sites, with the minimum of conflict between users. They provide ideal opportunities for establishing interpretive and educational centres through specialist voluntary organizations such as the Marine Conservation Society.

References

Addy, J.M. (1981). The macrobenthos of Sullom Voe. *Proceedings of the Royal Society of Edinburgh*, **80b**, 271–98.

Anderson, D.T. (1978). Cirral activity and feeding in the coral-inhabiting barnacle *Boscia anglicum*. *Journal of the Marine Biological Association of the United Kingdom*, **58**, 607–26.

Atkins, S.M. (1983). Contrasts in benthic community structure off the North Yorkshire coast (Sandsend Bay and Maw Wyke). *Oceanologica Acta*, Special Volume December 1983, 7–10.

Atkinson, R.J.A. and Nash, R.D.M. (1985). Burrows and their inhabitants. *Progress in Underwater Science*, **10**, 109–15.

Bamber, R.N. (1984). The benthos of a marine fly-ash dumping ground. *Journal of the Marine Biological Association of the United Kingdom*, **64**, 211–26.

Bennett, D.B. and Brown, C.G. (1983). Crab (*Cancer pagurus*) migrations in the English Channel. *Journal of the Marine Biological Association of the United Kingdom*, **63**, 371–98.

Bonsdorff, E. and Vahl, O. (1982). Food preference of the sea urchins *Echinus acutus* and *E. esculentus*. *Marine Behaviour and Physiology*, **8**, 243–8.

Brenchley, G.A. (1982). Mechanisms of spatial competition in marine soft-bottom communities. *Journal of Experimental Marine Biology and Ecology*, **60**, 17–33.

Briggs, J.C. (1974). *Marine Zoogeography*. McGraw-Hill, New York.

Broom, D.M. (1975). Aggregation behaviour of the brittle-star *Ophiothrix fragilis*. *Journal of the Marine Biological Association of the United Kingdom*, **55**, 191–7.

Bryan, G.W. (1976). Some aspects of heavy metal tolerance in aquatic organisms. In: *Effects of pollutants on aquatic organisms*, pp. 7–34. Lockwood, A.P.M. (ed.) Cambridge University Press, Cambridge.

Brylinsky, M. (1972). Steady-state sensitivity analysis of energy flow in a marine system. In: *Systems Analysis and Simulation in Ecology*, pp. 81–101. Patten, B.C. (ed.). Academic Press, New York.

Buchanan, J.B. (1966). The biology of *Echinocardium cordatum* (Echinodermata: Spatangoidea) from different habitats. *Journal of the Marine Biological Association of the United Kingdom*, **46**, 97–114.

Buchanan, J.B. and Warwick, R.M. (1974). An estimate of benthic macrofaunal production in the offshore mud of the Northumberland coast. *Journal of the Marine Biological Association of the United Kingdom*, **54**, 197–222.

Buchanan, J. B. and Moore, J. J. (1986). A broad review of variability and persistence in the Northumberland benthic fauna – 1971–85. *Journal of the Marine Biological Association of the United Kingdom*, **66**, 641–57.

Buchanan, J.B., Brachi, R., Christie, G. and Moore, J.J. (1986). An analysis of a stable period in the Northumberland benthic fauna – 1973–80. *Journal of the Marine Biological Association of the United Kingdom*, **66**, 659–70.

Caufield, C. (1984). More toxic waste to be burned at sea. *New Scientist*, **102**, 10.

Clark, R.B. (1986). *Marine Pollution*. Clarendon Press, Oxford.

Collins, K.J. and Mallinson, J.J. (1984). Colonisation of the 'Mary Rose' excavation. *Progress in Underwater Science*, **9**, 67–74.

Collins, N.R. and Williams, R. (1982). Zooplankton communities in the Bristol Channel and Severn Estuary. *Marine Ecology – Progress Series*, **9**, 1–11.

Cushing, D.H. (1980). European fisheries. *Marine Pollution Bulletin*, **11**, 311–15.

Davies, J.M., Addy, J.M., Blackman, R.A., Blanchard, J.R., Ferbrache, J.E., Moore, D.C., Somerville, H.J., Whitehead, A. and Wilkinson, T. (1984). Environmental effects of the use of oil-based drilling muds in the North Sea. *Marine Pollution Bulletin*, **15**, 363–70.

Dring, M.J. (1982). *The Biology of Marine Plants*. Edward Arnold, London.

Dyer, M.F., Fry, W.G., Fry, P.D. and Cranmer, G.J. (1982). A series of North Sea benthos surveys with trawl and headline camera. *Journal of the Marine Biological Association of the United Kingdom*, **62**, 297–313.

Dyer, M.F., Fry, W.G., Fry, P.D. and Cranmer, G.J. (1983). Benthic regions within the North Sea. *Journal of the Marine Biological Association of the United Kingdom*, **63**, 683–93.

Eagle, R.A. and Hardiman, P.A. (1976). Some observations on the relative abundance of species in a benthic community. In: *Biology of Benthic Organisms*, pp. 197–208. Keegan B.F., O'Ceidigh, P. and Boaden, P.J.S. (eds). Pergamon Press, Oxford.

Erwin, D.G. (1983). The community concept. In: *Sublittoral Ecology. The ecology of the shallow sublittoral benthos*, pp. 144–64. Earll, R. and Erwin, D.G. (eds). Clarendon Press, Oxford.

Farnham, W.F. and Bishop, G.M. (1985). Survey of the Fal Estuary, Cornwall. *Progress in Underwater Science*, **10**, 53–63.

Faubel, A., Hartig, E. and Thiel, H. (1983). On the ecology of the benthos of sublittoral sediments, Fladen Ground, North Sea. 1. Meiofauna standing stock and estimation of production. *Meteor 'Forschungsergebnisse'*, **36**, 35–48.

Fenchel, T. (1978). The ecology of micro- and meiobenthos. *Annual Review of Ecology and Systematics*, **9**, 99–121.

Forteath, G.N.R., Picken, G.B., Ralph, R. and Williams, J. (1982). Marine growth studies on the North Sea oil platform Montrose Alpha. *Marine Ecology - Progress Series*, **8**, 61–8.

Forteath, G.N.R., Picken, G.B. and Ralph, R. (1983). Interaction and competition for space between fouling organisms on the Beatrice Oil Platforms in the Moray Firth, North Sea. *International Biodeterioration Bulletin*, **19**, 45–52.

George, C.L. and Warwick, R.M. (1985). Annual macrofauna production in a hard-bottom reef community. *Journal of the Marine Biological Association of the United Kingdom*, **65**, 713–35.

Glemerec, M. (1973). The benthic communities of the European North Atlantic continental shelf. *Oceanography and Marine Biology. An Annual Review*, **11**, 263–89.

Goodman, K.S. and Ralph, R. (1981). Animal fouling on the Forties Platforms. In: *Marine Fouling of Offshore Structures*. Society for Underwater Technology, London, 19–20 May.

Gordon, J.C.D. (1983). Some notes on small kelp forest fish collected from *Saccorhiza polyschides* bulbs on the Isle of Cumbrae, Scotland. *Ophelia*, **22**, 173–83.

Gray, J.S. (1981). *The Ecology of Marine Sediments*. Cambridge University Press, Cambridge.

Griffiths, A. and Dennis, R. (1984). Adopt a site. Seven years of rock watching. *Sea* Marine Conservation Society Newsletter. March 1984, pp. 2–3.

Gulliksen, B. (1980). The macrobenthic rocky-bottom fauna of Borgenfjorden, North-Trondelag, Norway. *Sarsia*, **65**, 115–38.

Hartley, J.P. (1984). The benthic ecology of the Forties Oilfield (North Sea). *Journal of the Marine Biological Association of the United Kingdom*, **80**, 161–95.

Hepper, B.T. (1977). The fishery for crawfish, *Palinurus elephas*, off the coast of Cornwall. *Journal of Experimental Marine Biology and Ecology*, **64**, 251–9.

Hiscock, K. (1983). Water movement. In: *Sublittoral Ecology: The ecology of the shallow sublittoral benthos*, pp. 58–96. Earll, R. and Erwin, D.G. (eds). Clarendon Press, Oxford.

Hiscock, K. and Hoare, R. (1975). The ecology of sublittoral communities at Abereiddy Quarry, Pembrokeshire. *Journal of the Marine Biological Association of the United Kingdom*, **55**, 833-64.

Hiscock, K. and Mitchell, R. (1980). Description and classification of sublittoral epibenthic ecosystems. In: *The Shore Environment*, Vol. 2, *Ecosystems*, pp. 323-70. Price, J., Irvine, D.E.G. and Farnham, W. (eds). Systematics Association Special Volume **17b**.

Hoare, R. and Peattie, M.E. (1979). The sublittoral ecology of the Menai Strait. I. Temporal and spatial variation in the fauna and flora along a transect. *Estuarine and Coastal Marine Science*, **9**, 663-75.

Holme, N.A. and Wilson, J.B. (1985). Faunas associated with longitudinal furrows and sand ribbons in a sand-swept area in the English Channel. *Journal of the Marine Biological Association of the United Kingdom*, **65**, 1051-72.

Holme, N.A. and Rees, E.I.S. (1986). An interesting deep-water community in the Irish Sea. *The Challenger Society Newsletter*, **22**, 15.

Hughes, R.G. (1975). The distribution of epizoites on the hydroid *Nemertesia antennina* (L). *Journal of the Marine Biological Association of the United Kingdom*, **55**, 275-94.

Jensen, A.C. and Sheader, M. (1986). A description of the infauna present off Sellafield, N.E. Irish Sea, during May 1983. *Porcupine Newsletter*, Vol. 3, **8**, 193-200.

Jones, D.J. (1971). Ecological studies on macro-invertebrate communities associated with polluted kelp forest in the North Sea. *Helgolaender Wissenshaftliche Meeresuntersuchungen*, **22**, 417-41.

Jones, N.S. (1950). Marine bottom communities. *Biological Reviews*, **25**, 283-313.

Jones, N.S., and Kain, J.M. (1967). Subtidal algal colonization following the removal of *Echinus. Helgolaender Meeresuntersuchungen*, **15**, 460-6.

Kain, J.M. (1979). A review of the genus *Laminaria. Oceanography and Marine Biology: An Annual Review*, **17**, 101-61.

Kaplan, S.W. (1983). Intrasexual aggression in *Metridium senile. Biological Bulletin*, **165**, 416-18.

Kaplan, S.W. (1984). The association between the sea anemone *Metridium senile* (L) and the mussel *Mytilus edulis* (L) reduces predation by the starfish *Asterias forbesii* (Desor). *Journal of the Marine Biological Association of the United Kingdom*, **79**, 155-7.

Kennelly, S.J. and Underwood, A.J. (1984). Underwater microscope sampling of a sublittoral kelp community. *Journal of Experimental Marine Biology and Ecology*, **76**, 67-78.

Kitching, J.A., Macan, T.T. and Gilson, H.C. (1934). Studies in sublittoral ecology. I. A submarine gulley in Wembury Bay, S. Devon. *Journal of the Marine Biological Association of the United Kingdom*, **19**, 677-705.

Lack, T.J. and Johnson, D. (1985). Assessment of the biological effects of sewage sludge at a licensed site off Plymouth. *Marine Pollution Bulletin*, **16**, 147-152.

Lane, D.J.W., Nott, J.A. and Crisp, D.J. (1982). Enlarged stem glands in the foot of the post-larval mussel, *Mytilus edulis*: adaptation for bysso-pelagic migration. *Journal of the Marine Biological Association of the United Kingdom*, **62**, 809-18.

Lee, A.J. and Ramster, J.W. (eds) (1981). *Atlas of the Seas around the British Isles*. Ministry of Agriculture, Fisheries and Food.

Lilly, S.J., Sloane, J.F., Bassindale, R., Ebling, F.J. and Kitching, J.A. (1953). The ecology of the Lough Ine rapids with special reference to water currents. IV. The sedentary fauna of sublittoral boulders. *Journal of Animal Ecology*, **22**, 87-122.

Luning, K. (1971). Seasonal growth of *Laminaria hyperborea*. In: *Fourth European Marine Biology Symposium*, pp. 347-61. Crisp, D.J. (ed.).

Madin, L.P. (1985). *In situ* photography of gelatinous zooplankton. In: *Underwater*

Photography for Television and Scientists, pp. 65–82. George, J. D. Lythgoe, G. I. and Lythgoe, J. N. (eds). Underwater Association Special Volume No. 2. Clarendon Press, Oxford.

McCall, P.L. (1978). Spatial-temporal distributions of Long Island Sound infauna: the role of bottom disturbance in a nearshore marine habitat. In: *Estuarine Interactions*, pp. 191–219. Wiley, M.L. (ed.). Academic Press, New York.

McIntyre, A.D. and Eleftheriou, A. (1968). The bottom fauna of a flatfish nursery ground. *Journal of the Marine Biological Association of the United Kingdom*, **48**, 113–42.

McKenzie, J.D., and Moore, P.G. (1981). The microdistribution of animals associated with the bulbous holdfasts of *Saccorhiza polyschides* (Phaeophyta). *Ophelia* **20**, 201–13.

Manuel, R.L. (1981). *British Anthozoa*. Academic Press, London.

Milton, P. (1983). Biology of littoral blenniid fishes on the coast of south-west England. *Journal of the Marine Biological Association of the United Kingdom*, **63**, 223–37.

Moal, J., Samain, J.F., Koutsikopoulos, C., Le Coz, J.R. and Daniel, J.Y. (1985). Ushant thermal front: digestive enzymes and zooplankton production. In: *Proceedings of the nineteenth European Marine Biology Symposium*, pp. 145–56. Gibbs, P.E. (ed.). Cambridge University Press.

Moore, P.G. (1973). The kelp fauna of north-east Britain. I. Introduction and the physical environment. *Journal of Experimental Marine Biology and Ecology*, **13**, 97–125.

Newell, R.C., Field, J.G. and Griffiths, C.L. (1982). Energy balance and significance of micro-organisms in a kelp bed community. *Marine Ecology - Progress Series*, **8**, 103–13.

O'Donnell, A.R., and Mance, G. (1984). Estimation of the loads of some List I and List II substances to United Kingdom tidal waters – a comparison with previous estimates. *Water Pollution Control*, **83**, 554–61.

Pearson, T.H. and Eleftheriou, A. (1981). The benthic ecology of Sullom Voe. *Proceedings of the Royal Society of Edinburgh*, **80B**, 241–69.

Peattie, M.E. and Hoare, R. (1981). The sublittoral ecology of the Menai Strait, II. The sponge *Halichondria panicea* (Pallas) and its associated fauna. *Estuarine, Coastal and Shelf Science*, **13**, 621–35.

Petersen, C.G. (1913). Valuation of the sea. II. The animal communities of the sea bottom and their importance for marine zoogeography. *Report of the Danish Biological Station to the Board of Agriculture*, **21**, 1–44.

Pomeroy, L.R. (1980). Detritus and its role as a food source. In: *Fundementals of Aquatic Ecosystems*, pp. 84–102. Barnes, R.K. and Mann, K.H. (eds). Blackwell Scientific Publications, Oxford.

Potts, G.W. and McGuigan, K.M. (1986). A preliminary survey of the distribution of postlarval fish associated with inshore reefs and with special reference to *Gobiusculus flavescens* (Fabricius). *Progress in Underwater Science*, **11**, 15–25.

Ralph, R. (1986). A platform for fish? *New Scientist*, **109(1495)**, 42–4.

Ralph, R. and Goodman, K. (1979). Foul play beneath the waves. *New Scientist*, **82**, 1018–21.

Rees, H.L. (1983). Pollution investigations off the north-east coast of England: community structure, growth and production of benthic macrofauna. *Marine Environmental Research*, **9**, 61–110.

Rees, E.I.S., Nicholaidou, A. and Laskaridou, P. (1977). The effects of storms on the dynamics of shallow water benthic associations. In: *Biology of Benthic Organisms*, pp. 465–74. Keegan, B.F., O'Ceidigh, P. and Boaden, P.J.S. (eds). Pergamon Press, Oxford.

Riedl, R. (1964). Die Erscheinungen der Wasserbewegung und ihre Wirkung auf

Sedentarier im mediterranen Felslitoral. *Helgolaender Wissenschaftliche Meeresuntersuchungen*, **10**, 155–86.

Rivers, J.B., Pearson, J.E. and Schultz, C.D. (1972). Total and organic mercury in marine fish. *Bulletin of Environmental Contamination and Toxicology*, **8**, 257–66.

Robbins, I.J. (1985). Ascidian growth and survival at high inorganic particulate concentrations. *Marine Pollution Bulletin*, **16**, 365–7.

Robinson, G.A., Aiken, J. and Hunt, H.G. (1986). Synoptic surveys of the western English Channel. The relationships between plankton and hydrography. *Journal of the Marine Biological Association of the United Kingdom*, **66**, 201–18.

Rubin, J.A. (1980). Spatial interaction in a sublittoral benthic community. *Progress in Underwater Science*, **5**, 137–46.

Russell, F.S. (1935). On the value of certain plankton animals as indicators of water movements in the English Channel and North Sea. *Journal of the Marine Biological Association of the United Kingdom*, **20**, 309–22.

Schmidt, G.H. (1983). The hydroid *Tubularia larynx* causing 'bloom' of the ascidians *Ciona intestinalis* and *Ascidiella aspersa*. *Marine Ecology – Progress Series*, **12**, 103–5.

Schmidt, G.H. and Warner, G.F. (1984). Effects of caging on the development of a sessile epifaunal community. *Marine Ecology – Progress Series*, **15**, 251–63.

Shaw, D.F. (1983). Conservation and development of marine and coastal resources. In: *The Conservation and Development Programme for the UK*, pp. 261–313. Kogan Page, London.

Sheppard, C.R.C., Bellamy, D.J. and Sheppard, A.L.S. (1980). Study of the fauna inhabiting the holdfasts of *Laminaria hyperborea* (Gunn) Fosl. along some environmental and geographical gradients. *Marine Environmental Research*, **4**, 25–51.

Shin, P.K.S. (1981a). Temporal variations in a shallow subtidal community in North Bay (inner Galway Bay), west coast of Ireland. *Irish Naturalist Journal*, **20** (8), 3321–4.

Shin, P.K.S. (1981b). The development of sessile epifaunal communities in Kylesalia, Kilkieran Bay (west coast of Ireland). *Journal of Experimental Marine Biology and Ecology*, **54**, 97–111.

Sieburth, J., McN. and Conover, J.T. (1965). Sargassum tannin, an antibiotic which retards fouling. *Nature*, **208**, 52–3.

Stebbing, A.R.D. (1971). The epizoic fauna of *Flustra foliacea* (Bryozoa). *Journal of the Marine Biological Association of the United Kingdom*, **51**, 273–300.

Steel, J.H. (1975). *The structure of marine ecosystems*. Harvard University Press. Cambridge, Massachusetts.

Stuart, V. and Klumpp, D.W. (1984). Evidence for food-resource partitioning by kelp-bed filter feeders. *Marine Ecology – Progress Series*, **16**, 27–37.

Tebble, N. (1976). *British Bivalve Seashells*. HMSO, Edinburgh.

Tegner, M.J. and Levin, L.A. (1983). Spiny lobsters and sea urchins; analysis of a predator-prey interaction. *Journal of Experimental Marine Biology and Ecology*, **73**, 125.

Thompson, T.E. (1976). *Nudibranchs*. T.F.H. Publications Ltd, New Jersey.

Vanosmael, C., Willems, K.A., Claeys, D., Vincx, M. and Heip, C. (1982). Macrobenthos of a sublittoral sandbank in the southern Bight of the North Sea. *Journal of the Marine Biological Association of the United Kingdom*, **62**, 521–34.

Vogel, S. (1974). Current-induced flow through the sponge *Halichondria*. *Biological Bulletin of the Marine Biological Laboratory, Woods Hole*, **147**, 443–56.

Warner, G.F. (1971). On the ecology of a dense bed of the brittle-star *Ophiothrix fragilis*. *Journal of the Marine Biological Association of the United Kingdom*, **51**, 267–82.

Warner, G.F., (1979). Aggregation in echinoderms. In: *Biology and Systematics of*

Colonial Organisms, Systematics Association Special Volume No. 11, pp. 375–96. Larwood, G. and Rosen, B. R. (eds). Academic Press, London.

Warner, G.F. (1985). Dynamic stability in two contrasting epibenthic communities. In: *Proceedings of the nineteenth European Marine Biology Symposium*, pp. 401–10. Gibbs, P.E. (ed.) Cambridge University Press.

Warwick, R.M. (1984). The benthic ecology of the Bristol Channel. *Marine Pollution Bulletin*, **15**, 70–6.

Warwick, R.M., and Davies, J.R. (1977). The distribution of sublittoral macrofauna communities in the Bristol Channel in relation to the substrate. *Estuarine Coastal and Shelf Science*, **5**, 267–88.

Whittle, K.J. (1977). Marine organisms and their contribution to organic matter in the oceans. *Marine Chemistry*, **5**, 381–411.

Wicksten, M.K. (1983). Camouflage in marine invertebrates. *Oceanography and Marine Biology Annual Review*, **21**, 177.

Williams, R. (1986). Thermal structure and its effect on zooplankton communities. *The Challenger Society Newsletter*, **22**, 7.

Wilson, S.R. (1982). Horizontal and vertical density distribution of polychaete and cirrepede larvae over an inshore rock platform off Northumberland. *Journal of the Marine Biological Association of the United Kingdom*, **62**, 907–17.

Young, C.M. and Chia, F-S (1984). Microhabitat-associated variability in survival and growth of subtidal solitary ascidians during the first 21 days after settlement. *Marine Biology*, **81**, 61–8.

Yule, A.B. and Walker, G. (1985). Settlement of *Balanus balanoides*: the effect of cyprid antennular secretion. *Journal of the Marine Biological Association of the United Kingdom*, **65**, 707–12.

Index

Printed in the United States
By Bookmasters